The International Library of Ethics, Law and Technology

Volume 25

Technologies are developing faster and their impact is bigger than ever before. Synergies emerge between formerly independent technologies that trigger accelerated and unpredicted effects. Alongside these technological advances new ethical ideas and powerful moral ideologies have appeared which force us to consider the application of these emerging technologies. In attempting to navigate utopian and dystopian visions of the future, it becomes clear that technological progress and its moral quandaries call for new policies and legislative responses. Against this backdrop, this book series from Springer provides a forum for interdisciplinary discussion and normative analysis of emerging technologies that are likely to have a significant impact on the environment, society and/or humanity. These will include, but be no means limited to nanotechnology, neurotechnology, information technology, biotechnology, weapons and security technology, energy technology, and space-based technologies.

Eike-Henner W. Kluge

The Ethics of Artificial Intelligence in Healthcare: From Hands-on Practice to Policy-making

Springer

Eike-Henner W. Kluge
Department of Philosophy
University of Victoria
Victoria, BC, Canada

ISSN 1875-0044 ISSN 1875-0036 (electronic)
The International Library of Ethics, Law and Technology
ISBN 978-3-031-72836-5 ISBN 978-3-031-72837-2 (eBook)
https://doi.org/10.1007/978-3-031-72837-2

This Springer imprint is published by the registered company Springer Nature Switzerland AG
The registered company address is: Gewerbestrasse 11, 6330 Cham, Switzerland

Preface

Contemporary healthcare, whether at the societal, institutional or hands-on level, is the inheritor of tools, devices and techniques that have been developed over millennia. By and large, these have generally been perceived as involving parameters that are material in nature.

While this is correct, it is only partially correct because it ignores the fact that, in one way or another, the development of these tools, devices and techniques has always been data-dependent: i.e., that it has always been dependent on the amount and nature of the data that were available for their usage and for their design. This holds true not only at the level of actual practice and of health policy formulation but also at the level of health-related research and experimentation. In other words, in a very real sense the evolution of healthcare delivery has depended on something that has nothing to do with healthcare as such. To use contemporary terminology, it has depended on informatics in the wider sense of that term: on the science of data retrieval, storage and processing.

This science itself is and always has been multifactorial. The factors involved range from the material to the purely conceptual. Moreover, even in healthcare they also extended beyond the purely specific sphere and branched into other realms, as for instance into engineering and mathematics. Thus, with the advent of electronic methods of gathering data by using devices such as x-rays, sonographs and magnetic resonance imaging, the evaluation of the data that were thus gathered went hand-in-hand with developments of new mathematical approaches for analysing them because the traditional methods were no longer sufficient, and with the introduction of patient- rather than professionally originated data the mathematical models for analysing them required the development of still more robust mathematical means to evaluate their significance and to minimize or prevent interpretational error. That, however, was merely at the hands-on level. At the policy level, population-based data that dealt with the incidence and prevalence of particular health conditions became important, and here the informatic and mathematical methods for gathering and dealing with them underwent similar developments both in their nature and their sophistication.[1, 2]

One of the most important changes that occurred in this regard was the development of so-called Big Data and of artificial intelligence (AI). These have begun to play a crucial role both in the formulation of healthcare policies and in healthcare decision making at the institutional and the hands-on level.

The term "Big Data", of course, is really an acronym. It refers to huge data sets that are amassed using the modern technologies that were referred to a moment ago and that vastly transcend the amount of data that were available in the traditional healthcare setting. But, under the name of Bid Data Analytics,[3] the term Big Data is sometimes also used to refer to the methods and techniques that are used to analyse these huge data sets: methods that extract their significance in a short amount of time and that correlate them so that appropriate decisions can be made at the level of individual patients as well as at the policy level[4] in order to be able to deal with such things as the global incidence and prevalence of health issues. The situation of the COVID-19 outbreak of 2019[5] and of the Ebola outbreak of 2014,[6,7] when healthcare policy makers could not otherwise have reacted in a speedy fashion to fulfil their mandate of providing appropriate healthcare for their respective subject-populations in an effective manner, is here a case in point.

In a functional sense, the devices that engage in Big Data Analytics are AIs: they are computer programmes or expert systems. At the hands-on level, they analyse a given data set about a particular patient in order to arrive at a diagnosis when the corresponding analysis by a healthcare professional might transcend the professional's competence and go beyond their area of specialization or when doing so would be too time consuming. At the institutional level they are used to identify

---

[1] Newell B and Siri J. (2016). A role for low-order system dynamics models in urban health policy making. *Environment International* 95: 93–97.

[2] Physicians for a National Health Program. Healthcare Systems - Four Basic Models. Healthcare Systems - Four Basic Models | Physicians for a National Health Program. https://www.pnhp.org/single_payer_resources/health_care_systems_four_basic_models.php

[3] Ristevski B, Chen M. Big Data Analytics in Medicine and Healthcare. *J Integr Bioinform*. 2018 May 10;15(3):20170030.

[4] Khan ZF, Alotaibi SR. Applications of Artificial Intelligence and Big Data Analytics in m-Health: A Healthcare System Perspective. *J Healthc Eng*. 2020 Aug 30; Wang L, Alexander CA. Big data analytics in medical engineering and healthcare: methods, advances and challenges. *J Med Eng Technol*. 2020 Aug;44(6):267–283.

[5] Sharma A, Ahmad Farouk I, Lal SK. COVID-19: A Review on the Novel Coronavirus Disease Evolution, Transmission, Detection, Control and Prevention. *Viruses*. 2021 Jan 29;13(2):202; Imran M, Yasmeen R. SARS-CoV2 Outbreak: Emergence, transmission and clinical features of human coronaviruses. *J Ayub Med Coll Abbottabad*. 2020 Oct-Dec;32(Suppl 1)(4):S710–S713.

[6] Anema A, Kluberg S, Wilson K, et al. (2014). Digital surveillance for enhanced detection and response to outbreaks. *Lancet Infect Dis*. 14(11):1035–1037.

[7] Vayena E Salathé M, Madoff LC and Brownstein JS. (2015). Ethical challenges of big data in public health. *PLoS computational biology*, 11(2), e1003904. https://doi.org/10.1371/journal.pcbi.1003904

patterns and for devising protocols to deal with different types of situations in an institutional setting. They may even be constructed in such a way that they can correct the very algorithms that they themselves use to reach the conclusions that they do when these conclusions are subsequently found to be mistaken. The design of such expert systems has become increasingly sophisticated—so multifaceted, in fact, that the term "artificial intelligence" has been coined to reflect their sophistication—and it continues to advance at a steady pace.

However, this is only part of the story. Not only has the delivery of healthcare evolved with respect to the tools that it uses and the acquisition, handling and manipulation of data, it has also evolved with respect to the ethics of its delivery.

To put that into context, it may be appropriate to begin by considering the framework within which modern healthcare is delivered. Thus, for one reason or another—Article 25.1 of the *Universal Declaration of Human Rights* is usually cited in this regard[8]—it has become a more or less standard assumption in modern societies that everyone should have access to healthcare, whether that healthcare is construed on a commodity model or on a rights-oriented basis. However, since there is only a limited number of healthcare professionals and institutions that can deliver the relevant services, and because the individuals who require healthcare may reside in geographically distinct locations, the methods of delivering healthcare at the hands-on level have also expanded to include such things as eHealth or telemedicine—i.e., by engaging in healthcare consultations or providing diagnoses as well as actual hands-on healthcare interventions through the internet.[9,10] For these reasons, as well as for reasons of personal convenience, all of this is becoming increasingly common as the technology has increased in sophistication.[11] This increase in sophistication has even extended to the provision of hands-on healthcare by using machines whose use is initiated by the healthcare professionals but where the actual intervention is directed and performed by the AIs. Telesurgery is here a case in point.[12]

However, all of these developments have come with an ethical problematic that has not been resolved in a uniform and consistent fashion. This problematic involves not only the use of these methodologies and technologies but also their design, and it affects their hands-on users and institutional administrators as well the policy-makers of the societal healthcare systems in which the care is delivered. The purpose of the present study is to outline how this is the case with respect to the use of AIs, and to sketch an ethical framework for dealing with the situation. And since its focus is ethics, it will consider the legal provisions that different nations or groups

[8] United Nations, Universal Declaration of Human Rights, available at https://www.un.org/en/about-us/universal-declaration-of-human-rights

[9] Eysenbach G. What is e-health?. *J Med Internet Res*. 2001;3(2):E20.

[10] Jin, Michael X et al. "Telemedicine: Current Impact on the Future." *Cureus* vol. 12,8 e9891. 20 Aug. 2020.

[11] Park YT. Emerging New Era of Mobile Health Technologies. *Healthc Inform Res*. 2016;22(4):253–254.

[12] Choi PJ, Oskouian RJ, Tubbs RS. Telesurgery: Past, Present, and Future. *Cureus*. 2018;10(5):e2716. Published 2018 May 31.

of nations have enacted in this regard only when this is required for illustrative purposes.

By that very token, however, it will be theory- and even philosophy-heavy. This immediately invites the comment that then it has little relevance for the real world of actual healthcare, because the real world does not follow nice and tidy theoretical and philosophical lines of reasoning: It consists of facts that are messy, complicated and have little to do with theory. And as Christian Garve—a lecturer in philosophy at the University of Leipzig and a contemporary of Immanuel Kant—already said in his review of Kant's *Critique of Pure Reason*, "That may be fine in theory, but is unworkable practice."[13]

In reply, one can do no better than to paraphrase the answer that Kant himself gave in response to this criticism: If considerations of practicality are the only ones that are relevant to everyday life, then thinking about the nature of the world and of how it ought to proceed ethically has been reduced to nothing more than an intellectual parlour game or to something that university students debate in seminars and to which they listen in lectures.

In actuality, of course, that is not the case. The fact is that philosophical and ethical considerations have practical implications. This is particularly the case with ethics, as the *Universal Declaration of Human Rights* and the very existence of the International Court of Justice[14] and of international courts and tribunals that adjudicate on such things as war crimes and crimes against humanity illustrate only too well. In fact, ethical considerations are universally held to transcend national laws, and jurisdictional or legal provisions are ultimately considered to be subject to ethical considerations. Therefore a study such as the present one, which proposes to sketch an ethical framework for dealing with the use of artificial intelligence and expert systems in the planning and delivery of healthcare, is not a pointless exercise nor—even though it may at times be philosophy-heavy—is it out of touch with reality.

At the same time, embarking on such an undertaking imposes a series of requirements on its very conduct if it is to amount to more than merely a tacit appeal to accept its conclusions. It requires that the ethical approach that will be used in the course of the enquiry be clearly presented and justified, and that the technical terms that will be used will be clearly identified and explained. Consequently, since there are a variety of ethical theories or approaches, the first chapter of this study is partly devoted to identifying and justifying the particular ethical theory that will be used in the chapters that follow. And since the notions of artificial intelligence and of health can also be understood in a variety of ways, another part of the first chapter is devoted to explaining how and why they will be interpreted in a specific manner.

[13] Kant I." Über den Gemeinspruch: Das mag in der Theorie richtig sein, taugt aber nicht für die Praxis." *Zum Ewigen Frieden: Ein philosophischer Entwurf. Mit einer Einleitung* herausgegeben von Heiner F. Klemme. Meiner, Hamburg 1992. For an English translation, see Humphrey T, (ed. and transl.) *Immanuel Kant, Perpetual Peace and Other Essays* (Indianapolis: Hackett, 1983) "On the Common Saying: 'That may be correct in theory, but it is of no use in practice'.", pp. 61–92.

[14] International Court of Justice, https://www.icj-cij.org/en/court.

The same thing holds true about the notions of a healthcare professional and of a moral agent. Consequently, the second chapter is devoted to their clarification. The subsequent chapters will then engage in the actual analysis at the hands-on, institutional and social policy level.

Finally, a study such as this would be incomplete if it did not acknowledge that the constitution of AIs is subject to change, and that different ethical considerations become relevant as it changes. Thus, as they are currently constituted present-day AIs are merely expert systems and are merely tools. Therefore ethical considerations apply to them only indirectly precisely because they are tools. For, ethical considerations apply directly only to the persons who use tools. Arguably, however, this would change if the constitution of AIs changed and they became agents in their own right. That is to say, if they transitioned from being entities that functioned purely as tools to become entities that functioned in an independent manner and as persons. More specifically still, if they became what has sometimes been called artificial general intelligences (AGIs). The final chapter, therefore, will deal with this possibility. It is a possibility which, as quantum computing, bioprinting[15] and related technologies increase in sophistication, is neither far-fetched nor unlikely to happen.

Victoria, BC, Canada Eike-Henner W. Kluge

[15] Mandrycky C, Wang Z, Kim K, Kim DH. 3D bioprinting for engineering complex tissues. *Biotechnol Adv*. 2016 Jul-Aug;34(4):422–434; Matai I, Kaur G, Seyedsalehi A, McClinton A, Laurencin CT. Progress in 3D bioprinting technology for tissue/organ regenerative engineering. *Biomaterials*. 2020 Jan;226;

Contents

Chapter 1
Introduction: A Definitional Preamble

Abstract As its title indicates, this study deals with the ethics of artificial intelligences in healthcare. However, there are various ways in which the terms "artificial intelligence", "healthcare" or "ethics" may be understood. This first chapter, therefore, is devoted to outlining how they will be used in the chapters that follow and why they will be used in that way. Moreover, since there is a variety of ethical theories or approaches, it also indicates which one of these will be used in the analysis that is presented in the subsequent chapters, and why this approach is adopted.

Introduction

The term "healthcare" denotes the provision of health related care to individuals or to a community in order to deal with health related issues that negatively influence their physical, mental or emotional well-being. It may be provided on a non-professional basis—for instance, when it is provided by one family member to another or when it is self-administered—or it may be provided on professional basis; for instance, when it is provided by a physician, a nurse or a therapist. It may also be provided in an institutionalised setting such as a clinic, a hospital or by an agency. On the other hand, it may also be provided in an everyday setting such as the home of an individual or in another alternative setting. Finally, it may involve specialized items, tools or devices such as drugs, scalpels or ultrasound machines but it may also be provided simply in terms of person-to-person interaction.

In other words, the term "healthcare" may be understood in a variety of ways. However, no matter how it is understood, its purpose is to improve, maintain or ameliorate the health status of the recipient of the relevant provisions or actions. This, in turn, means that in order to be quite clear on what might be involved, the term "health" itself has to be clearly defined since otherwise the reason for engaging in the relevant actions or putting in place the relevant provisions remains uncertain.

The terms "artificial intelligence" and "ethics" can also be understood in a variety of ways. In their case, too, unless they are clearly defined, the subject of the

E.-H. W. Kluge, *The Ethics of Artificial Intelligence in Healthcare: From Hands-on Practice to Policy-making*, The International Library of Ethics, Law and Technology 25,
https://doi.org/10.1007/978-3-031-72837-2_1

present study—namely, the ethics of artificial intelligence in healthcare—will lack a precise focus. This initial chapter, therefore, is intended to provide the requisite clarity, and thereby to set the stage for the analysis that is presented in the subsequent chapters.

The Definition of "Health"

Beginning, then, with the definition of "health", the term was defined by the World Health Organization (WHO) in 1948 as "a state of complete physical, mental and social well-being and not merely the absence of disease or infirmity."[1] Because the WHO apparently considered this definition too imprecise, it restated it in 1958 as follows[2]

> …the extent to which an individual or group is able, on the one hand, to develop and satisfy needs; and, on the other hand, to change or cope with the environment. Health is therefore seen as a resource for everyday life, not the objective of living; it is seen as a positive concept emphasizing social and personal resources, as well as physical capacities.

However, this is not the only definition that has been offered. Thus, "health" has also been defined in statistical terms as the "statistically normal biological function with … typical efficiency" for a given species.[3] Still other definitions include "a person's psychophysiological ability to act or respond (in a way that is supportive of the person's goals, projects or aspirations) in a wide variety of situations;"[4] and "the ability to achieve … vital goals,"[5] as well as "a state characterized by anatomic, physiologic, and psychological integrity". It has even been defined as the "ability to perform personally valued family, work, and community roles" and "the ability to deal with physical, biological, psychological, and social stress."[6] It has also been defined in normative terms: for instance as the absence of defect or deviance, where how the notions of defect and deviance are understood depends on the philosophical view that one has of human nature.[7] Furthermore, it has been suggested that an

---

[1] *Constitution of the World Health Organization*, adopted in 1946 and entered into force in 1948; https://apps.who.int/gb/bd/PDF/bd47/EN/constitution-en.pdf

[2] World Health Organization (WHO). (1958).*Basic Documents: Constitution;* https://apps.who.int/gb/bd/index.html

[3] Boors C. (1977). "Health as a theoretical Concept," *Philosophy of Science* 44:4, 542–573.

[4] Whitbeck C: (1981). "A Theory of Health," in Caplan AL, Engenhardt HT Jr. and McCatrney JJ. (eds.) *Concepts of Health and Disease.* Adison-Wesley, 611–626 at. 620.

[5] Nordenfelt LY (1995). *On the nature of Health: An Action-Theoretic Approach*. Springer.

[6] Stokes, J.; Noren, J.; Shindell, S. (1982-01-01). "Definition of terms and concepts applicable to clinical preventive medicine". *Journal of Community Health.* 8 (1): 33–41.

[7] Engelhardt, Jr. HT. (1975). "The Concepts of Health and Disease," in *Evaluation and Explanation in the Biological Sciences* ed. Engelhardt and Spicker. Dordrecht: Reidel, 125–141.

appropriate definition should include parameters that span "the whole clinical field and would be applicable with respect to chronic illness and age-related phenomena."[8]

There are still other definitions.[9] However, the variety of these definitions notwithstanding, several things are important when it comes to considering the ethics of artificial intelligence in healthcare. Perhaps the most important of these, and as a matter of overarching requirement, is that the definition of "health" must be unequivocal if it is going to be used in the design of an artificial intelligence (AI). As will be explained more fully later, the reason for this is that AIs are computer programmes. Computer programmes, however, consist of instructions or algorithms that a computer will execute. These algorithms must not only be logically consistent with each other, they must also be precise. Consequently, if AIs are to be used in healthcare, the need for an unequivocal and non-circular definition of "health" is simply a matter of logic. Moreover, the definition must be valid in practical terms: that is to say, it must agree with the facts of reality. Otherwise, an AI that was structured by the relevant algorithms may well be able to function in a consistent fashion, but it would not be usable in the real world that deals with facts.

An Analysis

As to the various definitions that were mentioned above, they fail on various grounds. Thus—and focusing first on the definition that suggests that health should be understood in terms of statistically normal biological function—the fact that something is statistically normal in a given population or even in a given species does not necessarily entail that the population or the species is healthy in any robust sense of that term. Thus, nearly 95% of people in the United States have had an Epstein-Baer virus infection or have had mononucleosis by the time they are adults.[10] That, however, does not mean that Epstein-Barr virus infections or suffering from mononucleosis are not disorders and matters of ill health even though the majority of the population may suffer from it at any given time. Similar remarks apply to the common cold.

Moreover, a definition of health that simply centres on human beings would also be unacceptable because it would not take into account the fact that human beings are animals. In fact, they are social animals. That is why an appropriate definition has to take into account the biological functioning of their bodies as well as their cognitive, psychological and psycho-social functioning. Consequently, an

---

[8] Leonardi F. The Definition of Health: Towards New Perspectives. *Int J Health Serv.* 2018 Oct;48(4):735–748.

[9] McCartney G, Popham F, McMaster R, Cumbers A. Defining health and health inequalities. *Public Health.* 2019 Jul;172:22–30.

[10] Tyring, Stephen; Moore, Angela Yen; Lupi, Omar (2016). *Mucocutaneous Manifestations of Viral Diseases: An Illustrated Guide to Diagnosis and Management* (2 ed.). CRC Press. p. 123.

acceptable definition of health that could be used to design of an AI that was to be used in healthcare would also have to span these parameters.

In this respect, the amended WHO definition might seem to be useful. However, even that definition is not without its difficulties. For, the physiological, psychological and psycho-social state and functioning of human beings changes over time as they progress through different stages of their life cycle. Therefore what may be considered healthy at one stage of their development may not be what would be considered healthy at another. For instance, the pharmacokinetic functioning of a 2-year old human being typically differs from that of a 50-year old human being, and what is healthy in a prepubescent child is not necessarily what is healthy in an individual who is sexually mature. What is healthy also differs for individuals who, either by choice or by genetics, belong to different sexes—and even what the notion of sex itself amounts to. Thus, anatomy and hormones play a distinct role the health status of females as opposed to males.[11] Moreover, the definition does not address the issue of genetic disorders such as albinism, Klinefelter syndrome or Tourette syndrome. Therefore, a definition of health that does not take all of this into account would be incomplete, and an AI that was programmed in terms of such a definition would have limited usefulness at best.

Finally, the WHO definition—whether in its original or in its modified form—actually cannot be used in the design of an AI in healthcare for two other reasons: Thus, in its original formulation it is blatantly circular because it includes the notion of disease. That notion, however, is parasitic on the concept of health. Therefore an AI that was designed on the basis of this sort of definition might well result in endless and circular processing loops, and certainly would not be usable outside of the socio-economic context in which it was designed. As to its modified version, the ability to develop and satisfy needs ignores the fact that some needs—for instance, the need to take pornographic advantage of incompetent persons or to express a satyriasistic tendency—may constitute mental disorders and may even be identified as such in the *International Classification of Disease.*[12] Moreover, the ability to cope with the environment is often a matter of resources and not the result of a health condition.

Not only that: A usable definition of health also has to acknowledge that the notion of health is not specific to human beings—in other words, that it is not a species-specific notion—but that it applies to all biological beings. It is only in this way that it is usable in the design of AIs that can be used in hands-on healthcare as well as in the experimental protocols that are intended to assess possible healthcare interventions for human beings by testing them on non-human animals.

[11] Crimmins EM, Kim JK, Solé-Auró A. Gender differences in health: Results from SHARE, ELSA and HRS. *Eur J Public Health* 2010;21:81–91; Case A, Paxson C. Sex differences in morbidity and mortality. *Demography* 2005;42:189–214.

[12] *International Classification of Diseases* (ICD)-10, Code 52.8; https://www.icd10data.com/ICD10CM/Codes/F01-F99/F50-F59/F52-/F52.8

A Possible Definition

Therefore—and taking all of this into account—a possible of definition of health that would avoid these and similar issues would be that health is a state of homeostatic balance that provides an organism with the capacity to survive and function and to produce viable offsprings in the context in which it is embedded.

This definition not only is unambiguous, it also avoids the shortcomings of the other definitions that were mentioned above. Thus, it is not species-specific: As was said, humanity is not the only species that has health issues. Nor is it restricted to physical health. Further, it acknowledges that an entity may be healthy to various degrees, that genetics are an integral component of health, that what may be healthy in one context need not be healthy in another, and that an individual's ability to respond to environmental challenges may determine whether that entity is healthy. It also acknowledges that deviation from the norm is not necessarily the same as ill health. Moreover, it does not reduce to individual, social or conceptual standards or values while at the same time acknowledging that existing within a social context may have health-relevant features. Finally, it is consistent with the generally accepted position that someone may be healthy in one aspect of their being—for instance, with respect to their psychological condition—without being healthy with respect to another, as for instance with respect to their physical organ systems or even with respect to particular aspects of their overall nature.

When the notion of health is understood in this fashion, it spans a correspondingly wide array of services. More particularly, it covers all services whose purpose is to restore or maintain a functional state of health, or to at least ameliorate any insufficiencies that may be identified in functional terms. At the same time, it expands the field in which artificial intelligence may find application beyond the merely physical. Moreover, it suggests that but for a few exceptions[13] much of the current developments of AIs in healthcare have largely been confined to matters of physical well-being: That therefore they are too limited, because they tend to ignore the fact that human health may have social and mental dimensions. Furthermore, it also allows for the use of data that are not necessarily concerned with hands-on health issues but that deal with population health issues: for instance, those that are used in policy development. Likewise, it allows for the design and use of AIs in the design, implementation and evaluation of experimental protocols that include non-human animal where the results are then transferred to human beings. Finally, it is neutral on the issue whether a particular health condition automatically entails a mandate for treatment. Consequently, it leaves open the issue what role autonomy and informed consent play—or should play—in the domain of healthcare.

[13] Graham S, Depp C, Lee EE, Nebeker C, Tu X, Kim HC, Jeste DV. Artificial Intelligence for Mental Health and Mental Illnesses: an Overview. *Curr Psychiatry Rep*. 2019 Nov 7;21(11):116.

The Definition of Artificial Intelligence

Just like the term "health", the term "artificial intelligence" (AI) and its cognates also has a variety of uses and is understood in a variety of ways.

Putting it first in purely scientific terms, it designates a computer system that consists of many functionally integrated calculative routines and subroutines that handle data.[14] Common understanding, however, tends to go further: It frequently takes an AI to be a computer-based electronic system that mimics the problem-solving and decision-making capabilities of the human mind.[15] In this sense, it has been defined as a machine with the "ability to perform the cognitive functions we associate with human minds, such as perceiving, reasoning, learning, interacting with the environment, problem-solving, and even exercising creativity."[16] It has also been understood as a computer-based system that, once it has been activated, is able to operate independently and without continuous supervision or instruction by a human being. It has further been characterized as a system that can modify its own functional framework as it responds to new types of data, and that can adjust its operational subroutines in order to correct or prevent any shortcomings or mistakes that may have resulted from its operating according to its initial processing parameters. It is sometimes even been understood as a system that can decide on its own whether and how to respond to a given situation: In other words, that it not only has the capacity to learn but that it also possesses volition, just like a rational human being. The AIs that are portrayed in the 1984 movie *The Terminator* and in its sequels are presented as such systems. On this understanding, an artificial intelligence is what Stephen Hawking has called a "full artificial intelligence,"[17] or what John Searle has called a "strong artificial intelligence".[18]

In healthcare, an AI is often understood to be an electronic system that deals with patients on the basis of the data that are available to it. Using the tools that are available to it, it can evaluate a patient's health status, reach a diagnosis as well as make a prognosis. Moreover, rather than simply and automatically following the programmed instructions that are written into its routines and subroutines, it also has value-based parameters that allow it to choose among the possible tools for assessing a patient and for deciding among the various possible treatment modalities that it would present for the patient's or for their duly empowered substitute decision maker's approval. Sometimes it has even been understood as being able to change

---

[14] Russell and Norvig (2021). *Artificial Intelligence: A Modern Approach* (fourth ed.), Pearson, pp. 1–4.

[15] IBM, *What is artificial intelligence (AI)?* accessed 15/09/2023 at https://www.ibm.com/topics/artificial-intelligence; McKinsey & Company, *What is AI?* accessed 15/09/2023 at https://www.mckinsey.com/featured-insights/mckinsey-explainers/what-is-ai

[16] McKinsey & Company. (2024). "What is AI (artificial intelligence)?"accessed 04\05\2024 at https://www.mckinsey.com/featured-insights/mckinsey-explainers/what-is-ai

[17] Hawking S, (2018). *Brief Answers to the Big Questions*, Hodder & Stoughton.

[18] Searle J. (1984). *Minds, Brains and Science*: The 1984 Reith Lectures, Harvard University Press.

the valuational framework on the basis of which it had previously operated: More particularly, that it can change from using a purely professionally-centered medical value system to using a patient-centred value system instead. In other words, that it can modify the whole framework of its interactions with patients.

On this understanding as well, if it is connected to medical tools and devices it can perform hands-on interventions, and can be trained in this regard and can learn from its mistakes or shortcomings, just like its human counterpart. In other words, that it is what is sometimes called a learning system. It can even change its operational framework from purely locally focused actions to those that include geographically extended methods of operation such as telehealth, ehealth and teleconsulting. Furthermore, it can not only follow existing protocols for dealing with specific types of situations but can even initiate new ones if the existing ones are insufficient. In short, understood in this sense it functions much like a human healthcare professional except that it has an electronic basis.

Analysis

As was said, it is in this sense that the term "artificial intelligence" is frequently used not only in ordinary parlance but also in some of the contemporary scientific and technical discussions.

However, there currently is no artificial intelligence in this wider sense of that term—what in the remainder of this study will be called an AI-1. What does exist are AI-2 s or what will simply be called AIs: complex electronic systems that, in their calculations, can approximate the conclusions that an expert human professional would reach when faced with a particular type of situation, or reach conclusions of the sort that a human policy maker would reach on the basis of the data and in light of the resources that are available to them.[19] AIs as they currently exist, however, have neither awareness nor volition. They are merely complicated electronic systems that function according to how they are programmed, and they only function when their functioning is initiated by a human person. As such, they are used not only in the context of daily medical practice, telemedicine and telehealth[20] but also in identifying and evaluating possible protocol options at the institutional and policy level. They are also used in the context of medical research and in Big Data[21] in order to identify patterns, trends, and associations.

Once employed, these AIs can function without constant human input and, if so designed, can even refine their interpretational subroutines to reach ever more accurate conclusions as more data become available to them. Therefore, they can indeed

---

[19] Russell SJ and Norvig P. *op. cit.*

[20] Kuziemsky C, Maeder AJ, John O, et al. Role of Artificial Intelligence within the Telehealth Domain. *Yearb Med Inform.* 2019;28(1):35–40. https://doi.org/10.1055/s-0039-1677897.

[21] For an explanation of 'Big Data', see Preface p. i.

be designed as learning systems. Some AIs can even perform medical interventions such as surgery,[22] and when used by policy makers can be used to identify such things as the incidence and prevalence patterns of diseases and health conditions. As was said, however—and this bears emphasising—AIs as they currently exist have neither volition nor awareness, and they cannot function independently of being activated by the professionals who use them. Therefore, as was indicated before and unless specifically stated otherwise, in the current chapter and in the chapters that follow the term "artificial intelligence" will be understood in this the sense, and it will simply be referred to by the abbreviation "AI". Only in Chap. 6 will what were called AI-1 s be considered.

Ethics

There is just as wide an interpretation of the term "ethics" as there is for the term "health". For instance, it may be understood as referring to so-called teleological ethics, deontological ethics, communitarian ethics, virtue ethics, feminist ethics, religious ethics and agapistic ethics—to mention but a few of the more well-known types. In fact, there are various and importantly different ways in which the term may be construed even in the case of each of these different types of ethics. Therefore, an AI that was constructed according to one ethical perspective would yield different results in its operation than an AI that was constructed from another ethical perspective. And this, in turn, means that AIs whose focus was healthcare could not necessarily be used in all healthcare settings, for healthcare decision-making involves more than merely data: It also involves assigning significance or value to the data, and this is functionally related to the ethical system that is used in their design.

Religious Ethics and AI

For instance, religious ethics approaches are based on the religious tenets to that characterize the religion in question. Of course it is technically possible to design AIs in accordance with these religious tenets. They would then interpret the data that they are given and would function in accordance with the particular religiously oriented value-framework that was characteristic of the relevant religion. However, unless religious tenets were universally shared—and more importantly still, unless the fundamental tenets of the relevant belief system were considered to be

[22] Gumbs AA, Abu-Hilal M, Tsai TJ, Starker L, Chouillard E, Croner R. Keeping surgeons in the loop: are handheld robotics the best path towards more autonomous actions? (A comparison of complete vs. handheld robotic hepatectomy for colorectal liver metastases). *Art Int Surg* 2021;1:38–51. https://doi.org/10.20517/ais.2021.07.

objectively valid—the use of such an AI could not be transferred from one setting to another unless the different settings subscribed to the same religious belief system. Consequently, such an AI could not be used in settings that spanned distinct socio-cultural venues that included adherents of different religions.

However, even if a particular socio-cultural venue subscribed to the same religion, that would not mean that an AI that was designed in accordance with the tenets of that religion would yield *ethically* acceptable results. So, for instance, a religious belief system such as the one that is espoused by a fundamentalist Taliban society like Afghanistan deals with health issues that affect female persons on a different basis than those that affect male persons.[23] This, however, violates the ethical principles that are enunciated in *Universal Declaration of Human Rights* which stipulate that all human beings, insofar as they are human beings, are equal and should be treated in an equitable manner. This means that ethically it is irrelevant whether an individual is a male or female person: their health condition should be treated on the basis of their health status, and their gender is relevant only insofar as it affects their health.

Similarly, a fundamentalist Catholic Christian religious approach—and there are more than 600 hospitals and 1600 long-term care Catholic health facilities in the United States of America[24]—rejects a woman's right to a voluntary abortion and also rejects medical assistance in dying. Once more, therefore, if such a stance was integrated into the structure of an AI in healthcare it would limit the usability of that AI outside of that particular religiously oriented setting.

Perhaps even more significant is the fact that a religious ethics, no matter what the religion, is an ethics that is based on belief. However, the fact that something is believed does not entail that what is believed is true. In matters of ethics, just as in matters of mathematics or physics, the determining factor is not whether something is believed but whether it can be established on the basis of logic or of factually valid premises, where the reason that these are considered valid is not simply because they are believed, for that would be logically circular and therefore not probative. Therefore, while an AI whose subroutines are based on a religious perspective would function in a rule-like manner, it would not necessarily function in factually valid manner or yield factually valid conclusions. Consequently, as was already pointed, such an AI would be usable only with the particular religiously oriented healthcare setting.

---

[23] Rasekh Z, Bauer HM, Manos MM, Iacopino V. Women's health and human rights in Afghanistan. JAMA. 1998 Aug 5;280(5):449–55.

[24] Catholic Health Association, Facts and Statistics (2024), accessed 06/05/2024 at https://www.chausa.org/about/about/facts-statistics#:~:text=CHA%20advances%20the%20Catholic%20health,care%20providers%20in%20the%20nation

Virtue Ethics and AI

A similar thing holds true of virtue ethics. And, just as is the case with religious ethics, there is a variety of such ethics. For instance, the term "virtue ethics" may refer to Platonist virtue ethics,[25] exemplarist virtue ethics,[26] eudemonistic virtue ethics[27] and even to agent-based or feminist oriented[28] virtue ethics. What these all have in common, of course, is that the ethical considerations that fall into these various rubrics are not grounded in the facts of reality or on what are sometimes called fundamental ethical principles, such as those that are enunciated by the *Universal Declaration of Human Rights*. Instead, they are based on the set of virtues that define the particular system; and here a Platonic ethics would differ from an agent-based virtue ethics, as would a target-centred virtue ethics, etc.[29] And while a virtue-based ethics approach is easily implemented in the design and use of an AI—in other words, while it may easily be used to structure the value-gradients that are built into the AI's algorithms—it faces the problem of objective validity. Therefore any artificial intelligence whose overall valuational framework was designed on a virtue-ethics basis would also have limited usability.

More specifically still, on a virtue-ethics approach a particular action, protocol or policy is acknowledged to be virtuous if it is functionally determined by what is considered to be a virtue. However, as was just indicated, that differs from one virtue-ethical conceptual framework to another. For instance, the virtues that are central to Confucian ethics differ from those that are central to Stoic and Nietzschean ethics. Therefore a judgement that something is or is not virtuous is system-dependent and has no validity beyond the particular framework in which it is expressed. In other words it is, so to speak, functionally determined by the rules of the game.

However, the ethical the delivery of healthcare is not like playing a game. To be sure, it as well as the framework according to which it is structured should be rule-like and consistent. But merely being rule-like and consistent is insufficient to warrant the judgement that it is ethical. The rules themselves—the tenets around which the framework is structured—must be ethically valid.

Moreover, .as was said before, there is no universally agreed-on—let alone valid—set of virtues.[30] Therefore the framework of an AI that was structured in accordance with one set of virtues would differ from that of an AI that was structured in accordance with another. Consequently once more, just like a religiously

---

[25] Chappell T. (2014). *Knowing What to Do*, Oxford University Press.

[26] Zagzebski (2010). "Exemplarist Virtue Theory", *Metaphilosophy*, 41(1/2): 41–57.

[27] Badhwar NK. (2014). *Well-Being: Happiness in a Worthwhile Life*. Oxford University Press.

[28] Dillon RS. (2020). *Feminist Approaches to Virtue Ethics*. Oxford University Press.

[29] *Stanford Encyclopedia of Philosophy,* "Virtue Ethics", Sect. 3; https://plato.stanford.edu/entries/ethics-virtue/

[30] Nussbaum MC. (1999). Virtue Ethics: The Misleading Category *ARETÍ: revista de filosofía*, XI: 533–571.

grounded AI, it would be useless in healthcare that was provided in a diversified society, and would be of limited use at best.

Utilitarian Ethics and AI

A possible exception to some of the preceding considerations would seem to be an AI whose valuational framework was designed on the basis of a utilitarian ethics.

A utilitarian ethics is an ethics that is based on the Principle of Utility. That Principle stipulates that an action or rule of conduct is ethically appropriate if, all other things being equal, it produces or is likely to produce the greatest amount of good for the greatest number of people.[31,32]

The Principle itself is easily useable in structuring the relevant subroutines of an AI: for instance, by developing subroutines that calculate such things as quality adjusted life years (QALYs) or disability adjusted life years (DALYs). More specifically, a QALY calculation is simply a calculation of the utility-value of an outcome for an individual or for a group of individuals that is brought about by the treatment to which they are subjected when this is multiplied by the duration of the treatment.[33] QALYs can then be integrated with calculations of the medical costs of proceeding in that way so as to arrive at a final common denominator of Cost/QALY.[34] A DALY calculation is a calculation of the overall disease burden that is experienced by an individual or group of individuals as a result of suffering from a particular health condition. It is generally expressed as the number of years lost due to ill-health, disability or early death. In other words, mortality and morbidity are combined into a single common metric. Both types of measures are utilitarian in ethical orientation, and they are easily integrated into the calculative design of an AI. Moreover, both are often used in real life when evaluating the advisability of adopting a particular sort of healthcare intervention or of instituting a particular healthcare policy. In fact, they are used by the WHO[35] as well as by individual countries when assessing the acceptability of a particular type of healthcare intervention

[31] Brandt RB. (1979). *A Theory of the Good and the Right.* Oxford University Press.

[32] Stanford Encyclopedia of Philosophy, "The History of Utilitarianism," https://plato.stanford.edu/entries/utilitarianism-history/

[33] National Institute for Health Care and Excellence, "Judging whether public health interventions offer value for money," https://www.nice.org.uk/advice/lgb10

[34] Prieto L, Sacristán JA. (2003). Problems and solutions in calculating quality-adjusted life years (QALYs). *Health Qual Life Outcomes.* 19; 1:80.

[35] WHO, "Disability-Adjusted Life Years," https://www.who.int/data/gho/indicator-metadata-registry/imr-details/158#:~:text=DALYs%20for%20a%20disease%20or,%2C%20Sex%2C%20Cause%2C%20Risk%20factors

in terms of its outcomes, and to assist them in designing policies that maximize the health status of a particular societal setting.[36]

However, while presenting no real technical issues of construction, the use of an AI that was designed along these lines would be faced with a series of problems. Thus, whatever the particular version of utilitarianism may be—and there is as great a variety of utilitarianisms as there are of the other ethical theories that have already been mentioned[37]—the Principle of Utility itself is indeterminate with respect to what constitutes the greatest number of people and the temporal framework within which the relevant calculations should take place. Not only that, it also is indeterminate as to what counts as a good, and indeed how different types of goods should be evaluated relative to each other.

And while these issues could perhaps be resolved when designing the subroutines that structure an AI—if necessary, by simply deciding precisely how these parameters will be understood—the issue that could not thus be resolved, and what would ultimately rule out an AI that was designed from a utilitarian perspective from being usable in an ethically grounded healthcare system, is the fact that the Principle of Utility does not deal with persons insofar as they are persons: It treats them as objects.[38] It therefore completely ignores what is of central concern in ethics. Therefore, while an AI that was constructed according to the Principle of Utility might be usable in the practical planning and delivery of healthcare, its calculation would lack ethical validity.[39]

To show why this is the case, it may be appropriate to go into more detail by outlining how personhood and ethics are interrelated.

Personhood and Ethics

To put it first in general terms, ethics itself is the domain of morally right and wrong acts or actions. The notion of an act, however, is fundamentally distinct from that of an activity *per se*. An activity *per se*—or simply an activity—may be analysed purely in terms of the process that occurs when the activity is executed, whereas an act is something that not only involves the activity in question but also incorporates the volition of the actor. Thus, a bird that picks at a worm is engaged in an activity but, strictly speaking, is not engaged an act—except perhaps, metaphorically speaking. What it does is a matter of instinct, not of volition. Similarly, the knee jerk that

[36] LoomesG and McKenzie L. (1989). The use of QALYs in healthcare decision making. *Social Science & Medicine.* 28:4, 299–308.

[37] Lazari-Radek K. and Singer P. (2017). *Utilitarianism: A Very Short Introduction.* Oxford University Press.

[38] Rawls J. (1971). *A Theory of Justice.* Belknap Press; Rawls J. (2001). *Justice as Fairness: A Restatement.* Belknap Press.

[39] Prieto L, Sacristán JA. Problems and solutions in calculating quality-adjusted life years (QALYs). *Health Qual Life Outcomes.* 2003 Dec 19;1:80. https://doi.org/10.1186/1477-7525-1-80. PMID: 14687421; PMCID: PMC317370.

is produced when a reflex hammer triggers the muscle in the quadriceps of a human being is an activity, but it is not an act in the sense under discussion. It, too, is a reflex and not the result of the individual's decision. In other words, acts involve volition. The domain of ethics, therefore, is ultimately the domain of volition or of volition-associated factors. Since only persons are capable of volition, only persons can legitimately be the subjects of ethical considerations.

Moreover, volition or choice is possible only when there are options. If the situation in question does not present different options, one cannot choose. Not only that, the exercise of choice or volition requires that the agent must be aware of the options. One cannot choose between options if one is unaware of them. Furthermore, the agent must be free to choose among the options. An agent's lack of freedom to choose among available options entails that the agent is not a voluntary agent in that regard. That is why computer systems—and in particular AIs—may well select among different possibilities when responding to some input, but their doing so is not a matter of choice or volition but a matter of functional design.

Values

The matter goes further. Choosing among options requires values. Without values, the selection is either a random event or a happening that is predetermined by the nature of the entity in question. Moreover, all other things being equal—and this qualifier will be explained in a moment—the strength of the values determines which option is selected. It is in this sense that values are action potentials.[40]

That is to say, there are two distinct ways in which the term "values" can be understood: One refers to the causal reason that a distinct activity is performed when there are distinct possibilities. On this understanding, the reason that a particular possibility is selected is functionally related to the strength of the values in question. An example of this would be the values that distinct magnetic fields have with respect to each other: In each case, the field that has the greatest strength—differently, that has the strongest value—will predominate in its attraction. With due alteration of detail, this is also the way that values are incorporated into the design of a computer programme or AI. Thus, on this interpretation of "value", value is the significance that is attached to particular possibility—in other words, its logical nature—and the strength of that significance or value is the likelihood that the AI will make a particular selection among the various possibilities that are open to it at that point in time. It is, so to speak, like a vector in mathematical calculations. The selection, however, is neither a matter of cognition nor of volition. It is an automatic matter that is functionally determined by the design of the routines or subroutines in question.

---

[40] Hodgkinson, C. (1996). *Administrative Philosophy: Values and Motivations in Administrative Life*, Pergamon.

The other interpretation of "values" accepts that while values are action potentials, they are not determinative. Instead, acting on these values involves volition. Therefore, while their strength is functionally related to the likelihood that an agent will select a particular option or possibility when the relevant situation presents itself, the agent can choose whether or not they will act in accordance with these values. An illustration of this would be the case where a physician has a strong value that treats human life as precious but they nevertheless disregard that value in particular cases and provide medical assistance in dying.

Not only that: the individual who holds a given value has control not only over whether they will make a decision in keeping with that value but also over which values or action potentials they will adopt. It is in this sense that voluntary agents have values and make choices on the basis of their values. Computer programmes do not have values in this sense of the term. Consequently, AIs do not qualify as voluntary agents.

Personhood

Applying these considerations to the concept of a person, it means that persons are entities that have the capacity for cognitive awareness and volition and that can adopt, at their discretion, a particular set of values to guide their decision-making.

At this juncture, several other points become ethically important. The first is that none of this entails anything about the degree or acuity of cognitive awareness that an alleged person must have; nor does it entail that the capacity must in fact be actualized. Thus, persons do not cease to be persons when they are not actually cognitively aware or have little if any sophisticated understanding.

The second important point is that there are different kinds of persons because there are different kinds of entities that have the capacity for cognitive awareness and volition. These entities may be individually distinct—for instance, they may be individual members of the species *homo sapiens*—or they may be groups of individuals, as for instance the administrative boards of healthcare institutions whose members are collectively are aware of certain options and make a decision on a collective basis. The first kind of person is generally called a natural person, whereas the second kind of person is generally called a collective, constructive or corporate person.[41]

A third important point is that while all persons are voluntary agents, this does not entail that every choice they make is a voluntary matter and involves the application of values. They may simply react in a given situation.

The standard case of a person, of course, is that of an individual living member of the species *homo sapiens* who has the present capacity for cognitive awareness and volition. This in turn means that human beings who have not yet acquired that

[41] Davoudi L, McKenna C and Olegario R (2018). "The historical role of the corporation in society". *Journal of the British Academy*. 6: 17–47.

capacity—for instance, 12 week old fetuses—or human beings who no longer have that capacity—for instance, individuals who meet the Harvard Criteria for brain death[42] but are biologically alive—would not count as persons even though they would be living members of the species *homo sapiens*.

However, as was said, groups of individuals may function as a unit, and as such may be aware of certain facts, have certain values and aims, and make decisions on that basis. This is formally recognized ethically, socially as well as legally in the concept of a constructive or corporate person.[43] Like human persons, corporate persons have rights and duties. That it is why their actions can legitimately be evaluated using the same ethical principles as those that apply to individual human persons, and why both ethically and legally they can be held accountable for what they do—a point that will be discussed more fully in Chaps. 4 and 5 when dealing with institutions and policy-makers. Further, social groupings such as indigenous societies may also count as persons because as units they may be aware of certain facts, have certain values and make decisions on that basis. This becomes important when it comes to evaluating the use of AIs by healthcare institutions and agencies.

Finally, an issue that is beginning to assume some relevance at the present time is the question whether a human being ceases to be a person when their parts have been replaced by non-biological components. The matter is fairly clear in the case of external prosthetic devices such as arms or legs, and even in the case of internal prosthetic devices such as artificial mechanical hearts. These, after all, are not the basis for the individual's capacity for cognitive awareness and volition. However, as medical science progresses it is increasingly becoming possible to replace the biological components of human beings that ground their capacity for cognitive awareness and volition—in other words, their brain cells—with electronic chips.[44] The issue then becomes whether such individuals would still count as persons or whether they would count as AIs.[45]

Historically speaking, the issue had already been raised by Plutarch[46] and Hume[47] in general terms when they asked whether an object would remain the same object when its parts are replaced by other parts: for instance, when they considered the case of a ship whose parts were successively replaced by other parts. Their analyses, however, involved the issue of continued identity when the constituent parts of an

---

[42] "A definition of irreversible coma: report of the Ad Hoc Committee of the Harvard Medical School to Examine the Definition of Brain Death". *JAMA*. 205 (6): 337–40. 1968; https://doi.org/10.1001/jama.1968.03140320031009.

[43] Blair MM. (2013). Corporate Personhood and the Corporate Persona. *University of Illinois Law Review* 785.

[44] Shaima M et al. (2024). Elon Musk's Neuralink Brain Chip: A Review on 'Brain-Reading' Device. *Journal of Computer Science and Technology Studies* 6(1):200–203.

[45] Maguire GQ Jr., McGee EM. Implantable brain chips? Time for debate. *Hastings Cent Rep*. 1999 Jan-Feb;29(1):7–13; Holstein MB. Ethics and Alzheimer's disease: widening the lens. *J Clin Ethics*. 1998 Spring;9(1):13–22. PMID: 9598429.

[46] Plutarch, *Life of Theseus*, http://classics.mit.edu/Plutarch/theseus.html

[47] Hume D. *A Treatise of Human Nature* I: IV, II.

entity were replaced by materially equivalent parts of the same kind, whereas the emerging issue is whether the individual becomes a different kind of entity when their constituent parts are replaced by functionally equivalent parts that have a different material nature.

Arguably, and if the preceding analysis is correct, such human beings would still qualify as persons. The reason is that, purely logically speaking, what is central to being a person is not how their capacity for cognitive awareness and volition is grounded but whether they have that capacity. With due alteration of detail, this issue was already acknowledged when the notion of personhood was shown to include constructive or corporate persons. Therefore, if chip-replaced human beings retain the capacity for cognitive awareness and volition, they would still count as persons.

However—and this may be thought to undermine the preceding reasoning—chips, whether bioprinted or purely mechanical in nature, are subject to the laws of nature. Being purely material objects—and this includes ChatGPTs, which are cloud computing platforms—they have neither cognitive awareness nor volition. Therefore, unless it can be shown that the capacity for cognitive awareness and volition requires a dualistic metaphysics, a chip-replaced human being will no longer be a person. The matter will be discussed further in Chap. 6, whose subject is the use of AI-1 s or Artificial Generative Intelligences in the provision and planning of healthcare.

Implications

As was suggested in the initial parts of this discussion, however, there currently are no AI-1 s. There only are AIs, and these are tools that are used by persons. And while ethical considerations are indeed relevant when dealing with AIs, they are only indirectly applicable to the AIs themselves, and they can legitimately be called ethical or unethical only indirectly and only insofar as they are handled by their users. In and by themselves they are merely things. An example from outside of the sphere of healthcare and of artificial intelligence may serve to illustrate the point. An atomic bomb has incredible destructive potential, and an assault rifle will kill a human being—as will potassium cyanide and a hangman's noose. That, however, does not make these things unethical in themselves. They are unethical only in a derivative sense, and only when they are used in an ethically inappropriate manner.

Deontological Ethics and AI

It remains to outline the nature of deontological ethics and to indicate why it will be used in the subsequent chapters when considering the ethics of artificial intelligence in healthcare.

Deontological ethics—the term is ultimately derived from the Greek words "deon" or duty and "logos" or study—is the normative ethical theory that the ethical quality of an action and the ethical acceptability of the values that structure the framework of an individual's actions should be evaluated on the basis of certain fundamental principles rather than by simply focusing on the utility or outcome of the actions in question, and where these principles do not have a religious basis and are not grounded in virtues. These principles, in turn, are generally identified as the Principle of Autonomy, the Principle of Equality, the Principles of Beneficence and Nonmaleficence, the Principle of Fidelity, and the Principle of Impossibility respectively, and are reflected in the *Universal Declaration of Human Rights*[48] and in similar documents.[49]

In light of what was said before, the question that immediately arises is whether—and indeed how—such Principles could be claimed to have objective validity and not simply be reflective of a particular perspective. Also why they, rather than some version of the Principle of Utility, should be used when considering the ethics of artificial intelligence in healthcare.

The answer to these questions is most easily given by considering the Principles themselves and how they are derived.

Justification of Using Deontological Ethics

Since, as was pointed out above, the domain of ethics is ultimately the domain of volition and of volition-associated factors such as acts, and since (as was also pointed out) volition entails the ability to choose, which in turn entails autonomy, the appropriate Principle with which to begin is the Principle of Autonomy.

The Principle of Autonomy

An initial statement of this Principle is as follows.

> *Every person has the right to self-determination subject only to the equal and competing right of other persons.*

This Principle finds its basis in logic: Since persons are entities that have the capacity for cognitive awareness and volition, the very exercise of their nature entails that they can act as voluntary agents when the situation either allows or calls for their doing so. They need not do so, of course, but they may do so if they so wish. Moreover, except in a sense that will be explained later—substitute decision-making here enters the picture—engaging in volition is not something that one person can

---

[48] United Nations, *Universal Declaration of Human Rights*, https://www.un.org/en/udhrbook/pdf/udhr_booklet_en_web.pdf

[49] *Stanford Encyclopedia of Philosophy*, Deontological Ethics, https://plato.stanford.edu/entries/ethics-deontological/

do for another person but that they can only do for themselves. Therefore it logically follows that by their very nature a person—i.e., a voluntary agent—is a self-determining agent. In other words, it is simply a different way of saying that they are a person. Therefore the Principle of Autonomy is grounded in the very nature of persons insofar as they are persons.[50]

The basis of the limiting condition, in turn, is found in another Principle of deontological ethics: namely, the *Principle of Equality*:

The Principle of Equality

Since all persons insofar as they are persons are the same then, logically speaking, they should be treated the same way unless that there is something different relative to their natures as individuals or in their respective situations. In deontological ethics, this feature is captured in the Principle of Equality, and it may be stated as follows:

> *All persons, insofar as they are persons, are equal and should be treated the same. Exceptions must always be based on ethically relevant differences in their respective natures or condition.*

Consequently—and this explains the limiting clause in the Principle of Autonomy—if the right to self-determination of one person conflicts with the right to self-determination of another person, then it is logically impossible to accord the same status to their respective claims for autonomous action because their respective right- or autonomy-claims cannot met be at the same time. This, in turn, requires a resolution if deontological ethics is to be workable in practical terms.

And here—as will be discussed in greater detail in Chap. 4 when discussing the use of AIs by healthcare administrators—several possibilities present themselves. Thus, it is possible to argue that in such a case one individual's autonomy should simply be ranked lower than that of the other. However, unless there was an ethically relevant reason for doing so, that would violate the Principle of Equality itself.

Another way of resolving the impasses might be to rank the conflicting claims in terms of the nature of rights that are claimed. More specifically, it is arguable that some right are more fundamental than others, and on that basis claim that the exercise of a more fundamental right takes priority over the exercise of a less fundamental right. To illustrate the point at issue, it may be useful to put it in terms that were used by a noted American legal scholar some years ago, "Your right to swing your arms ends just where the other man's nose begins."[51] In other words, the right to personal integrity is more fundamental than the right to freedom of movement.

[50] Cf. *Stanford Encyclopedia of Philosophy*, "Autonomy in Moral and Political Philosophy", https://plato.stanford.edu/entries/autonomy-moral/

[51] Zechariah Chafee, Jr. (1919). "Freedom of Speech in War Time" 32 *Harvard Law Review* 932, 957.

Therefore proceeding in this way would resolve the conflict in question without denying the ethical equality of the respective claimants themselves.

As to when there is no fundamental difference between the competing right-claims, then either both rights would lapse or—as will be explained more fully in Chaps. 4 and 5 when dealing with allocation of resources—each individual should be assigned an equal chance and lots should determine whose right would be accorded priority. Nor is this merely a matter of arbitrary decision-making. Since each person would be treated equally, it is grounded in logic. It also finds reflection in international pronouncements[52] and law.[53]

Clearly—and as will be spelled out further in the succeeding chapters—both the Principle of Autonomy and the Principle of Equality can be incorporated into the design and use of AIs in healthcare. All that would be necessary is for the AIs' algorithms to check for competing right-claims, alert their users if that was the case, and to suggest relevant treatment parameters to deal with the situation. Of course that means that the algorithms would have to contain gradients that allowed them to calculate whether a particular right-claim was more fundamental than another. However, as will also be discussed more fully in Chap. 4, that does not present an insuperable problem in the logical design of an AI.

The Principle of Impossibility

Another deontological Principle—and one that also has profound implications for the construction and use of AIs in healthcare—is the Principle of Impossibility:

> *All rights and duties hold subject to the condition that it is possible to meet them under the circumstances that obtain, where the fact of impossibility lies in the nature of the rights or duties or of the situation and is not a result of the voluntary actions of the relevant party in question.*

The fundamental nature of this Principle is recognized not only in ethics but also in law. Thus, it already was an integral determinant of Roman law,[54] and it is fundamental to modern jurisprudence. To cite but a few specific examples, it finds reflection in the laws of India,[55] Japan[56] and Thailand,[57] and it clearly was fundamental to the leading case of *Taylor v. Caldwell* in the United Kingdom.[58] It also is generally

[52] *Universal Declaration of Human Rights*, Article 1.

[53] Petrova, D., 'The Declaration of Principles on Equality: A Contribution to International Human Rights Law', *Equal Rights Review*, Volume 2, December 2008, pp58–72.

[54] Justinian, *Digest* 50. 18. 185.

[55] India, *Contract Act* 1972 s. 56.

[56] Japan, *Civil Code* art. 415 is generally interpreted as having this as an underlying principle.

[57] See Chen-Wishart, Loke and Ong (eds.). *Studies in the Contract Law of Asia* (Oxford U Press, 2016) pp. 373, 391.

[58] *Taylor v Caldwell* [1863] EWHC QB J1.

accepted in international law even when evaluating the performance of treaty obligations.[59]

However, it is important to note that, ethically speaking, impossibility is not the same as futility. Futility is a question of pointlessness. Thus, while it would not necessarily be impossible to keep alive an individual whose higher brain centres are permanently dysfunctional—one of the longest cases in this regard lasted for 37 years[60]—it would be futile to do so in the sense that the individual would never again have sapient cognitive awareness. Therefore a properly constructed AI's design would incorporate the distinction between impossibility and futility if it was to indicate whether a given intervention was appropriate.

Ultimately, of course, the Principle of Impossibility is also grounded in logic. Thus, while it possible to articulate in a grammatically unexceptional manner and in one and the same expression rights and duties that logically conflict with each other, the fact that these would involve a logical contradiction entails that doing so would be logical nonsense and therefore would be conceptually meaningless. A famous example of this is the so-called Barber Paradox: In a little town there is barber who shaves all and only those who don't shave themselves: Who, then, shaves the barber?[61] While it is linguistically unexceptional, this characterisation of the barber is conceptually meaningful only with respect to the individual words.

In healthcare—as will be spelled out more fully in the later chapters—this Principle expresses itself in the fact that there is an obligation to provide care only if it possible to do so within the limits of the conditions that obtain. However, these limits must not be the result of the failure on part of the otherwise responsible agent to fulfil their obligation. For instance, a healthcare professional's obligation to employ a diagnostic tool such as an AI or an expert system lapses if such a tool is unavailable to the professional because it has not been developed or because the professional's institutional setting does not have the funds to provide it. However, if the professional cannot use the tool because they themselves have made it inoperative, then they are responsible for the impossibility and the professional is ethically guilty of not fulfilling their obligation. Likewise, the designer of an AI or of an expert system cannot be held morally responsible for not designing it to function at a level that exceeds the current level of technological sophistication.

The Principle of Fidelity

Another deontological Principle is that of Fidelity. It is also grounded in logic. It is to the effect that

---

59 See Reuter P. *Introduction au droit de traités* (Paris, 1985), pp. 152–157.

60 "Elaine Esposito dies after 37 years in coma". *Lewiston Evening Journal.* 27 November 1978.

61 Russell B. "The Philosophy of Logical Atomism," *Collected Papers of Bertrand Russell, 1914–19, Vol* 8., p. 228.

Whoever has an obligation has a corresponding duty to try to the best of their ability to meet that obligation.[62]

The reason it is grounded in logic lies in the very nature of a duty or obligation. More specifically, a duty or obligation is a requirement to do or to refrain from doing something. That requirement, however, is not fulfilled if the relevant action is not carried out at all, if it is carried out only partially, or if the requirement to refrain from doing something is not adhered to at all or is only partially adhered to. Thus, if the designer of an AI who was hired to develop an AI or expert system whose function was to engage in deep learning but did not use the best algorithms that were available to them, then passing off the AI as fulfilling the conditions of their employment would be to violate this Principle. Likewise, a healthcare administrator or healthcare professional who did not use the best AI or expert system that was available to them would also violate the Principle of Fidelity.

At the same time, however, this does not mean that all failures to meet an obligation fully—in other words, the failure to successfully achieve everything that is specified in the obligation—always constitutes an ethical default. Whether someone can fulfil their obligations depends on the circumstances in which they find themselves. If the circumstances are such that they make it impossible to meet the obligation, then—by the Principle of Impossibility—the obligation lapses.

This becomes relevant in the context of privacy and security considerations insofar as AIs or expert systems are concerned. Their operation is electronic, and involves the electronic storage and manipulation of data. However, as Shannon already proved in 1949, it is impossible to guarantee privacy unless one uses a Vernam cipher or one-time pad[63]—which of course is impossible when dealing with AIs or expert systems in healthcare. On the other hand, it would constitute an ethical fault if the designer of an AI did not take the greatest possible care to employ the best measures that are available to them under the circumstances that obtain where, as was pointed out above, relevant limiting circumstances range from lack of resources to the non-existence of the necessary technology. However, as will be explained more fully in succeeding chapters, it would be obligatory for the relevant designer, developer or user to acquaint whoever employed them with the existence and nature of such an impossibility or limitation. The duty to do so would follow from the Principle of Autonomy as well as the Principle of Fidelity. For, unless the employers were aware of such a limitation, they could not be said to have given informed consent for the design or development in the first place.

[62] Where an unstated but understood condition is that, ethically speaking, one cannot have an unethical obligation. This condition becomes relevant when distinguishing between ethically valid and purely contractual obligations.

[63] Shannon, C. (1949). "Communication Theory of Secrecy Systems". *Bell System Technical Journal* 28 (4): 656–715. https://doi.org/10.1002/j.1538-7305.1949.tb00928.

The Principles of Beneficence and of Nonmaleficence

Two other deontological Principles are the Principle of Beneficence and the Principle of Nonmaleficence. These Principles are correlatives of each other and, respectively, go as follows:

> *Everyone has a duty to advance the good of others insofar it is possible to do so without undue risk to oneself, where the nature of the good is in keeping with the competent values of the recipients of the action in question.*

and

> *Everyone has a duty to prevent harm insofar as this is possible without undue risk to oneself, where the nature of the harm is in keeping with the competent values of the recipient of the action in question.*

They are grounded in matters of logic and of fact. More specifically, persons insofar as they are persons—and this applies even to constructive or corporate persons and to collectivities—depend for their existence and functioning on society. Without society—without social interactions, however tenuous—they would not come into existence in the first place and they would not be able to survive and acquire those skills and characteristics that turn them into persons in the full sense of that term. Already Aristotle recognized this when—albeit with reference to biological persons—he defined a person as a *zoon politikon*: i.e., as a social animal.[64] With due alteration of detail, similar considerations apply in the case of corporate persons or collectivities.

In other words, personhood involves social embedding which, as was said, entails the existence of society. Since societies themselves are complexes of persons, this means that they logically requires the existence of other persons. That condition, however, cannot be met unless appropriate steps are taken to ensure the existence of the other persons *as person*—i.e., as autonomous individuals who have the capacity for cognitive awareness and volition. Therefore, since a society is a complex of persons and cannot engage in acts except through the agency of the persons who are its members, it means that relevant actions to this effect must be performed by the persons who are its members. Consequently, it follows that actions whose purpose is to underwrite the existence of other persons is a requirement for being or becoming a person. Differently, it means that if a person engages in an act, that act must contribute to or at least not endanger the existence of other persons *qua* persons. Otherwise the society, which is a complex of persons, would not be able to exist. That is why all members of society have a duty of beneficence and nonmaleficence.

However, as was pointed out above, persons *qua* persons are autonomous agents. Autonomy, in turn, amounts to self-determination. Since autonomy involves voluntarily making decisions, and since such decisions are made on the basis of values,[65]

[64] Aristotle, *Politics*, Book 1, section 1253a.

[65] Hodgkinson, *op. cit.*

it follows that the exercise of autonomy is conditioned by values. Equality entails that all persons, insofar as they are persons, are equal. Consequently all persons, insofar as they are persons, have the right to exercise their autonomy on the basis of their own values. From this, in turn, it follows that the other-directed acts of members of society—i.e., of persons—should not violate the values of the recipient of the action. Consequently, the values that are followed by someone who fulfils the duty of beneficence or nonmaleficence should not be autocratically determined by the actor on the basis of their own values. Instead, they should respect the values of the recipient of their action: i.e., of the other person. This explains the limiting subsidiary clause in the definition of the Principles of Beneficence and Nonmaleficence above.

However, since values are action potentials[66] and these vary in strength, depending on their nature and the situation in question, it also means that unless the relevant value gradients are integrated into the design of an AI, an AI could not be used to evaluate whether a given intervention or protocol was consistent with the Principles that were indicated above. Chapter 3 will explore how this is possible in the case of hands-on healthcare.

Conclusion

As was said in the beginning, the claim that ethical considerations are relevant to the design and use of AIs in healthcare is meaningful only if the terms that are used in presenting these considerations are clearly defined and the framework within which these considerations are adduced is clearly outlined and justified. The focus of the present discussion has therefore been to clearly outline the meaning of terms "health", "ethics" and "artificial intelligence" as they will be used in what follows.

An extended section has been devoted to justifying the use of deontological ethics as providing the criteria by which the design and use of AIs in healthcare should be measured. However, while deontological ethics will be treated as providing the ultimate measures of what is ethically appropriate in this regard, this does not mean that this is the only framework that will be considered in the chapters that follow. The use and implications of utilitarian ethics will also be considered. The reason for doing this is that healthcare, whether at the hands-on, the institutional or the policy level, is frequently developed using what essentially are utilitarian criteria. The healthcare systems that exist in Russia, the People's Republic of China as well as in India are good examples in this regard. Consequently, it seems appropriate to indicate how—and why—such an approach ultimately miscarries when considered from an ethical perspective. As to why the other ethical approaches that were mentioned above will not be treated the same way since they are also used in the design and delivery of healthcare—countries like Iran and Turkey essentially use

[66] Cf. P. 11, *supra*.

religiously based ethics—the answer is that this would require an analysis of how the logic of the relevant systems could be applied to the design and use of AIs. Since currently there is no available examination of how that would be the case, it would require an analysis in its own right; and that would transcend the limits of the present investigation.

At the same time, it may well be asked whether the various principles that should be followed in the design and usage of AIs and that have been enunciated by such organizations as the World Health Organization—for instance in its *Ethics and Governance of Artificial Intelligence for Health*[67]—or by the Organization for Economic Cooperation and Development—for instance, in its *Recommendation of the Council on Artificial Intelligence*[68]—would not have sufficed for outlining the ethical framework within which the design and usage of AIs in healthcare should proceed, rather than proceeding as above: Namely, by outlining and justifying the ethical system that will be adopted in the subsequent chapters, and by attuning the considerations that follow to the fundamental principles that structure this system.

The answer is very simple: While, undeniably, such pronouncements and recommendations have ethical content, they tend to focus more on the practical implications of the principles that they enunciate and do not provide insight into how these principles themselves are related to—and indeed are justified—in fundamental ethical terms. In other words, while they appeal to ethics, they do not show how ethics is involved at a fundamental level. Without that, however, while such guidelines and recommendations may have political and even legal import, they are subject to interpretation, and ultimately may be perceived to lack ethically binding force. The purpose of this study, however, is to indicate precisely what and how certain rules for dealing with AIs in healthcare are in fact ethically grounded, and why they apply irrespective of culture and jurisdiction. And that cannot be done without engaging in fundamental ethical considerations.

At the same time, this introductory discussion is incomplete as it stands because it presents and outlines only part of the conceptual framework of the present study. Other concepts that are involved include such things as to who precisely counts as a healthcare professional; and although this may seem to be clear in itself, this is not quite the case and has ethical implications for the use of AIs. Likewise, while the notion of personhood and the concept of a person were already discussed, they were dealt with only in the context of ethical theory and did not address the issue of how the concepts apply in the context of a professional agency and how they are relevant for the construction, design and use of artificial intelligences in healthcare—in particular with respect to AI-1 s as opposed to AI-2 s. However, dealing with this requires detailed argumentation, and that is more appropriately handled in a separate chapter. The next chapter, therefore, is devoted to this task.

---

[67] World Health Organization (WHO) Ethics and Governance of Artificial Intelligence for Health. Geneva, 2021, https://apps.who.int/iris/bitstream/handle/10665/341996/9789240029200-eng.pdf

[68] Organization for Economic Cooperation and Development (OECD), *Recommendation of the Council on Artificial Intelligence* 2019, https://legalinstruments.oecd.org/en/instruments/OECD-LEGAL-0449

Chapter 2
Healthcare Professionals, AIs, Tools and Persons

Abstract This chapter expands on the definitions that were presented in Chap. 1. Its focus is the different way in which the notion of a healthcare professional can be understood and the distinct ethical considerations that this entails. It also provides a more detailed explanation of the concept of a person than was given in Chap. 1, and indicates how this is relevant for the design and construction of AIs.

Introduction

In order to present a meaningful analysis of the relationship between AIs and healthcare professionals, it is appropriate to spell out more clearly what is understood by the relevant terms than was done in Chap. 1—in particular, by the terms "healthcare professional", "person" and "AI". For, it is only when these terms are clearly and precisely defined that the ethical considerations that are advanced in the rest of this study will amount to more than merely superficial reflections. The initial section of this chapter, therefore, will be dedicated to attempting such a clarification and to making the usage of these terms more precise. The remainder of the chapter will be devoted to examining in what sense AIs differ from a human healthcare professional in more than a purely material sense.

Healthcare Professionals

Beginning, then, with the term "healthcare professional," it is usually understood as referring to someone who is trained and engaged in the delivery of health care in a professional capacity.

However, this in itself is insufficiently precise because the domain of individuals who are trained and engaged in the delivery of healthcare in a professional manner includes a wide variety of individuals not all of whom have the same sort of training

E.-H. W. Kluge, *The Ethics of Artificial Intelligence in Healthcare: From Hands-on Practice to Policy-making*, The International Library of Ethics, Law and Technology 25,
https://doi.org/10.1007/978-3-031-72837-2_2

or are engaged in delivering healthcare on the same basis. Nor do they have the same a type of professional relationship with the actual recipients of that care—i.e., with patients. For instance, nurses differ in this regard from family physicians, surgeons, psychiatrist and radiologists; and even within these respective types of professions there are subgroups with distinct specialisations, and each of them relates to patients in a different manner. Nor, for that matter, do all of them use AIs in the same way.

Also, not all healthcare professionals of the same type provide the same array of services even though they may belong to the same professional rubric and may be similarly trained and qualified. Thus, Cristian nuns and monks who are trained and qualified as healthcare professionals and who work in the same areas as those who do not belong to such religious orders will not provide certain types of services such as medical assistance in dying, nor will they engage in abortions unless that can be justified in terms of the Doctrine of Double Effect.[1] This not only has ethical implications in itself, but also affects the nature and programming of AIs that they use in providing their services.

Likewise, not all healthcare professionals provide their services for the same reason. This not only may have ethical implications with respect to nature of the services that they provide but also with respect to programming of the AIs that they use when carrying out their profession. To revert back to the previous example, Christian nuns and monks who are professionally trained in healthcare provide their services not for financial but for religious reasons. So do Buddhist monks who are engaged in the hands-on delivery of healthcare. By contrast, other types of healthcare professionals generally have a pecuniary interest in doing what they do; and this not only has ethical implications in itself but also affects the quality and programming of the AIs that they use. After all, AIs are expensive items, and the cost for them has to be regained from the fees that these other professionals charge for providing their services. Therefore financial parameters constitute valid factors in the programming of their AIs.[2]

But even this comes with variations that have ethical relevance with respect to the nature and programming of the AIs that are used. More specifically, while it generally is the case that a pecuniary factor is the reason why many healthcare professionals engage in their professional activities, the professionals may voluntarily forego receiving such recompense from their patients. Therefore a fee structure—that is to say, financial parameters—may not be involved in the programming of the AIs that they use. For instance, this is the case with hands-on healthcare professionals who belong to such organizations as *Médecins Sans Frontières* (*Doctors Without Borders*) and whose work is underwritten by such organizations. While this would not affect the nature of their physician-patient relationship, it may nevertheless affect the quality and programming of the AIs that such professionals use. For, even

[1] *Stanford Encyclopedia of Philosophy*, "Doctrine of Double Effect". https://plato.stanford.edu/entries/double-effect/

[2] For a similar considerations, see Chap. 4, p. 66.

though there is no patient-focused financial factor in the provision of their services, the types of services that they provide and the limits within which they provide them is functionally related to the financial resources of the organizations that underwrite their activities. And this, of course, ultimately affects the nature and programming of the AIs that they use.

As an aside, all of this of course raises the question whether, when AIs turn out to be what in Chap. 1 were called strong AIs or AI-1 s, such AIs would also count as a healthcare professional. And in particular, it would also raise the question whether they then would also have the right to receive a recompense for their labour. And if so, it would raise the further question what the nature of that recompense would or should be.

However, abstracting from these questions—they will be addressed in the last chapter—an immediate answer to the question of who counts as a healthcare professional is that all individuals count as healthcare professionals if they function in the capacity of providing healthcare services on a professional basis. This, then, excludes individuals who engage in healthcare not for professional reasons but on some other basis. Therefore the ethical considerations that apply to the former do not necessarily apply to the latter nor, even if they did apply, would they apply in the same way.

At the same time, however, while it entails that the same overall ethical framework applies to all who work in a professional capacity in healthcare, how that framework applies—in other words, how the Principles of Ethics apply—varies from type of professional to type of professional.[3] It also means that how the Principles apply depends on the particulars of a given case. Therefore it may vary from case to case. An analogy may serve to illustrate the point: The fundamental laws of physics hold for the whole of the material universe and apply to everything that is contained in it. This means that they apply to the Moon as much as they apply to aardvarks. However, how much gravity an entity exerts is a function of the type of body that it has, and this is functionally related to the entity's mass.

Moreover, it raises the further question whether an AI that is not a strong AI but that functions independently once it is employed is also subject to ethical considerations. Put differently, it raises the further question whether such an AI would be the tool of the person who uses it or whether it would be an agent. This is ethically important because, as was pointed out in Chap. 1,[4] tools are subject to ethical evaluation only derivatively and not directly, whereas ethical considerations apply directly to the users of tools who qualify as agents and as persons.

In order to properly deal with this matter, it may be appropriate to take a closer look at the distinction between tools, agents and persons.

---

[3] This matter will be explored further in Chap. 3.

[4] Chapter 1, p.14.

Tools, Agents and Persons

By definition, a tool is a device or implement that is used to carry out a particular function or activity, whereas an agent is someone who may decide to use a tool in the performance of some function or action.[5] In other words, a tool is a mere object and as such has neither volition nor awareness, whereas an agent is an individual who is more than a mere object but is a person and has volition and awareness. Therefore, if the discussion above and in Chap. 1 is cogent, AIs are mere tools. That is why ethical considerations do not apply to them directly even when they carry out their function independently once they are engaged. It is only the persons or agents who employ or use them—in this case, the healthcare professionals—to whom ethical considerations apply directly.

At the same time, however, it might be argued that this inference is somewhat premature because, speaking purely logically, the same factors seem to be involved when a healthcare professional assigns a particular task to another individual. For instance, a surgeon who performs a cardiac bypass operation may tell an attending surgical colleague or a surgical resident to tie off certain peripheral blood vessels that have been cut when opening the patient's chest, or to keep open the patient's thoracic cavity while the surgeon busies themselves with the particulars of the patient's heart and its blood vessels. However, much if not all of such subsidiary tasks can also be carried out by AIs when they are sufficiently technically advanced.[6] Therefore it may well be asked whether, by parity of reasoning, such AIs should not also be considered persons. Or, alternatively, it may be suggested that the surgical colleague or resident should be considered to be tools: Human tools, to be sure, but tools nevertheless. Consequently, it may legitimately be asked in what sense—if at all—the two sorts of cases differ with respect to ethical considerations.

By way of reply, it is tempting to argue that an AI, even if it was acting independently once it was activated, would still be considered a tool simply because of its non-biological nature, whereas the colleague or the resident would always be considered agents because they are members of the species *homo sapiens* and have a biological nature.

It is also tempting to argue quite independently of this that AIs do not satisfy what are the defining characteristics of personhood: namely, cognitive awareness and volition. They simply are programmes in machines. Nor do they function independently of being activated. To be sure, they might function independently once they are activated: for instance, when they are activated by the surgeon in the example above. Also, they are unable to choose whether to carry out the relevant activity

---

[5] *Oxford English Dictionary*, "tool"

[6] Ishikawa N, Watanabe G. Robot-assisted cardiac surgery. *Ann Thorac Cardiovasc Surg*. 2015; 21(4):322–8; Rimmer L, Howard C, Picca L, Bashir M. The automaton as a surgeon: the future of artificial intelligence in emergency and general surgery. *Eur J Trauma Emerg Surg*. 2021 Jun;47(3):757–762. doi: 10.1007/s00068-020-01444-8. Epub 2020 Jul 26. PMID: 32715331; Moglia A, Georgiou K, Georgiou E, Satava RM, Cuschieri A. A systematic review on artificial intelligence in robot-assisted surgery. *Int J Surg*. 2021 Nov;95:106151.

once they are activated. As well, their activity—i.e., what they do—is functionally determined by the nature of their design. That design may include subroutines that allow them to modify their initial programming. However, even this is a matter of programming or design, and it is not a matter of choice on their part.

By contrast, so it might be argued, both the surgical colleague and the resident differ in this regard. Thus, whether they do anything—and even whether they are in the position in which they find themselves, such as being a surgical colleague or a resident—is a matter of choice on their part. By contrast, AIs have no such choice. Not only that: what the former do depends on how they understand the instructions that they are given, and therefore is functionally related to their awareness. Nor is what they do functionally determined by their design. They are not designed. It is functionally related to their understanding and training, which are not a matter of design. AIs differ in all of these respects.

In order to evaluate these considerations and to see whether they are correct, it may be beneficial to clarify what is involved in saying that a person has the capacity for cognition or cognitive awareness and volition whereas an AI does not, because this has profound implications with respect to an AI's ethical status: implications that go beyond what has been outlined so far. Moreover, it will also become relevant in the discussion of Chap. 6, which deals with AIs as genuine agents.

Cognition or Cognitive Awareness

Focusing first on the criterion of cognition cognitive awareness, as it is usually understood this notion refers to the ability of an individual to consciously determine the significance of what they are presented with.[7]

That statement, however, still leaves the issue of the nature of awareness unresolved because it leaves open the question of what is implicated by the modifier "consciously". In other words, it merely transfers the issue of how to understand the distinguishing characteristic of cognitive awareness to the question of what constitutes consciousness. After all, an entity that functions solely by instinct is also able to determine the significance of what it is presented with, but it does not therefore have conscious awareness.[8] Thus, a robin that selects only dry grasses for building its nest clearly determines the significance of what it is visually presented with, or a sea turtle that seeks out the ocean after hatching on the beach clearly determines the difference between the dry shore and the ocean. However, they can hardly be said to be consciously aware of the differences in question. They act solely on the basis of instinct. Therefore, unless the characteristic of consciousness has an aspect that is not found in instinct or in the processing systems of AIs, there is no reason to assume

[7] *Cambridge English Dictionary*, "cognitive: connected with thinking or conscious mental processes".

[8] *Ibid*, "instinct: a natural ability that helps you decide what to do or how to act without thinking"

that an AI that is performing certain surgical functions does not also possess the attribute or quality of cognitive awareness: In other words, that it is not also cognitively aware. The key point, then, is what the attribute of consciousness amounts to and how it is connected with the attribute of awareness.

Consciousness

At this juncture, it may be appropriate to widen the discussion and note that while human beings may also on occasion satisfy the criterion for being aware in the sense of determining the significance of that with which they are presented, that does not necessarily mean that when they do so, they do so in a cognitive mode; that is to say, that they do so in a cognitive and consciously aware manner. Thus, while premature babies *in utero* satisfy the criterion of consciousness or awareness insofar as they react to painful or other stimuli,[9] they cannot be said to be aware of these stimuli in a cognitively conscious manner. Cognitive awareness itself only develops later, when they have been born and have acquired concepts.[10] Moreover, even when they have acquired concepts—for instance, when they have acquired the concept of food—these concepts merely function as means for distinguishing between different types of sensory experiences. Initially they do not involve cognitive parameters in the full-blooded consciously-aware sense of that term. Thus, while a half-year old infant can tell the difference between milk and yoghurt, it cannot tell the difference between these *as* milk and *as* yoghurt. It is a mere matter of distinguishing. In that sense, therefore, they function much like AIs that effectively draw distinctions on the basis of the categories that are contained in their programming.

Furthermore, while a half-year old infant can tell the difference between themselves and milk and yoghurt, this does not mean that they are consciously and cognitively aware of themselves as not being milk or yoghurt. As psychological studies have shown, conscious cognitive self-awareness develops only gradually.[11] Not only that, automatous individuals may act in complex, co-ordinated and apparently purposeful ways and thus determine the significance—even the cognitive significance—of that with which they are presented. However, they do so without being consciously aware of doing so and therefore without being volitionally responsible for what they are doing.[12] This also is similar to what an AI does when it responds to stimuli. Just like the baby and the autonomous individual, it lacks conscious cognitive awareness.

---

[9] Lecanuet JP, Schaal B. Fetal sensory competencies. *Eur J Obstet Gynecol Reprod Biol.* 1996 Sep;68(1–2):1–23.

[10] Lagercrantz H, Changeux JP. The emergence of human consciousness: from fetal to neonatal life. *Pediatr Res.* 2009 Mar;65(3):255–60. doi: 10.1203/PDR.0b013e3181973b0d. PMID: 19092726.

[11] Jabr F. (2012). "Self-Awareness with a Simple Brain". *Scientific American Mind.* 23 (5): 28–29.

[12] Oyebode, F. (2018). *Sims' Symptoms in the Mind: Textbook of Descriptive Psychopathology* E-Book. Elsevier Health Sciences; "Automatism".

In other words, what seems to be missing in cases such as that of early stages of childhood development and of automatism, and what seems to transform cognitive awareness into conscious cognitive awareness or cognitive awareness in the full-blooded sense of that term, is the presence of a self-component in such awareness. It is the presence of self-awareness that seems to be integral to the characteristic of cognitive consciousness and that changes the nature of pure cognitive awareness to conscious cognitive awareness.[13] Clearly, if this is correct, it has important implications for the question whether AIs can ever be considered persons. For, if they cannot ever be consciously and cognitively aware in this sense, then this is not the case.

Moreover, both ethically and legally speaking, it seems to be this component of self-awareness that makes all the difference in how the behaviour of an agent is evaluated. Legally, this is captured by the fact that an automatous individual will not be held responsible for whatever actions they commit because they lack self-awareness while doing so.[14] As to ethically, already Kant, in his *Critique of Pure Reason*[15]and in his *Foundations of the Metaphysics of Morals*[16] insisted that this component is an integral factor of awareness insofar as ethical considerations apply: Further, that ethical considerations apply only to individuals who in that sense are aware (or who could reasonably be expected to be aware) of what they are doing. It is only when the individual's capacity for cognitive awareness in that sense is actualised that ethical considerations apply to their actions.

Volition

Analogous considerations apply in the case of volition. More specifically, both legally and ethically the notion has two distinct components: one that centres in the matter of being the causal originator of an activity, and one that centres in the attribute of intent. Thus, the fact that an individual is the causal originator of an activity or action is not itself sufficient to ascribe volition to them. The individual must intend the relevant activity or action. And this once more pushes the analysis back one step further: namely, to what is involved in the characteristic intending to do something or of having an intention or intent.

Clearly, of course—and except in a metaphorical sense[17]—one cannot intend something of which one is unaware. Moreover, as John Searle already pointed out in 1991, the having of an intent has an ineluctable awareness component, where this

[13] Lou HC, Changeux JP, Rosenstand A. Towards a cognitive neuroscience of self-awareness. *Neurosci Biobehav Rev.* 2017 Dec;83:765–773.

[14] Beran RG. Automatism: comparison of common law and civil law approaches—a search for the optimal. *J Law Med.* 2002 Aug;10(1):61–8. PMID: 12242885.

[15] Kant I. (1781). *Critique of Pure Reason* (A96)

[16] Kant I. (1785). *Foundations of the Metaphysics of Morals.*

[17] Hamlyn DW. Unconscious Intentions. *Philosophy* (London), 1971–01, Vol.46 (175), p.12–22.

component encompasses self-awareness.[18] Therefore the notion of volition, which but is another way of talking about the overall characteristic that grounds the having of an intent, includes the parameter of self-awareness. Not, to be sure, in the sense of the self being the focus of the awareness but in the sense of being the framework of that awareness. Consequently, an entity that has the capacity for volition must be an entity that has self-awareness: i.e., it must be an entity that not merely reacts but acts in a conscious and cognitively significant manner in the sense that was just outlined. It is worth emphasising that this does not necessarily mean that they need be aware of themselves as a self. However, it is only when the capacity for conscious cognitive awareness inclusive of self-awareness in the sense that was just outlined together with the capacity for volition is present in an individual, that ethical judgements can relate to them.

When these considerations are applied to the issue under discussion, then a fundamental difference between an AI that is used as a tool and a colleague or a resident who carries out an assigned task becomes apparent. If the colleague or resident at the bedside is a human being—even a human being whose bodily parts have been replaced by mechanical devices or parts of whose brain have been replaced by chips—then, in order to be subject to ethical considerations, they will have to exhibit cognitive awareness (inclusive of self-awareness in the above sense) as well as the capacity for making voluntary decisions. Only then will they be subject to ethical evaluation. And by the same token, if the AI at the bedside is an entity that has cognitive awareness in the sense that was just discussed as well as the capacity for making voluntary decision then—irrespective of whether it is constituted of electronic or biological components—that AI will also be subject to ethical considerations, just like the surgical colleague or resident.[19]

The Issue of Nature and Origin

At this juncture it may be argued that the origin of the entities in question—i.e., of the physician, of the physician's colleague, of the medical resident and of the AI—makes all the difference. For, unlike the former, an AI is constructed as a purely material entity and is non-biological in nature, whereas colleagues and residents have a gestational origin and (initially at least) have a purely biological in nature.

However, once again, that may not be entirely decisive. Brain cells are also material entities. Therefore they are also subject to the laws of nature. Arguably, therefore, the difference between having a biological and non-biological constitution does not make a fundamental difference.

[18] Searle J. "Consciousness, Unconsciousness and Intentionality," *Philosophical Issues* I (1991) 45–66.

[19] For a literary exploration of this issue, see Asimov I. (1946). "Evidence". *Astounding Science Fiction* vol. 9.

Nor—logically speaking—would the matter of origin make any difference. For, the professionals would have originated from fertilized ova and would have developed more or less on the basis of the genetic programmes that are contained in their DNA. In other words, they would not be what could be called self-originators. Therefore they would not differ from AIs in this regard, which also are not self-originators. In other words, both types of beings would have developed or been constructed on the basis of programmes that have been provided by their respective developers—and progenitors also fall into that category. After all, DNA is essentially a programme, and it differs from the programmes that underlie AIs only in that it is written in biological form. Furthermore, the fact that each of the human progenitors contributed only an incomplete genetic code—the gametes that give rise to a fetus are haploid—finds a parallel in the fact that the programmes that structure AIs also tend to have more than one contributor.

As to the manner or process of how the respective parties originate—so to speak, the manner of their construction—it could reasonably be argued that this would not make a fundamental difference either. For, once more, both the human professionals and the AIs are material in nature. This, in turn, means that in both cases how their respective codes are translated into different systems and ultimately how the complete entities are constructed is in accordance with the laws of nature. In both cases, it is also influenced by their respective environments and the resources that these environments provide. The only way, therefore—so the argument continues—in which the issue of origination could justifiably be said to ground a difference between the human professionals and AIs would be if at some stage in the development of human professionals their bodies acquired a non-material entity that was distinct from their brains, and that it was this entity that constituted the centre of their self-awareness and volition, whereas nothing like this was the case in the construction of AIs. In other words, that a fundamental difference in the constitution of human beings and AIs is that a dualistic metaphysics is necessary to explain the former but is not necessary to explain the latter.

Some Philosophical Considerations

There are historical precedents for making a dualistic claim in the case of human beings. For instance, this was proposed by Aristotle[20] and St. Augustine,[21] as well as by St. Tomas Aquinas[22] who enunciated the thesis of what came to be known as the

[20] Aristotle, *On the Generation of Animals* 735a 15–26 and 736a 35–736b.

[21] St. Augustine, *De spiritu et anima* 9 (PL 40, 784–85 ex); *Quaestiones Vetere Testamento* 23 (PL 35, 2229)

[22] St. Thomas Aquinas, *Commentary on the Sentences* (2, d. 18, q. 2, a. 3), *Summa contra gentiles* (2, 87–89), *Summa theologica* (1, q. 118, a. 2, ad 2.

thesis of "delayed hominization".[23] This was then further spelled out as the thesis that a soul can enter into or associate with a human body only after that body has developed an organ that is sufficiently complex to be able to sustain consciousness and volition: namely, a brain.

Historical antecedence, however, does not amount to proof, nor does even the internal consistency of a theoretical argument establish that what is maintained is in fact correct. Moreover, one could also argue that in order to be considered acceptable, the distinction between human persons and AIs should not only be in keeping with the facts, but that it should also explains the known facts in the simplest possible terms: In other words, that it should meet the criterion of what has become known as the Principle of Parsimony or of Occam's razor.[24]

However, even the last would not settle the point at issue. For, even if the simplest possible analysis that was consistent with the known facts satisfied the Principle of Parsimony, the facts that are known are not exhaustive of all the facts that there are. This is amply demonstrated by advances in the modern sciences. Moreover, strictly speaking, the Principle of Parsimony is not a logical principle at all but is heuristic. Therefore an appeal to parsimony may well persuade, but it does not establish.

Consequently, unless it can be shown that human persons have minds that are capable of doing something that AIs in principle cannot do, and unless it can be shown that the reason for this is not that the AIs are insufficiently developed in a technical sense but that this is necessarily the case, the difference between the human persons and AIs is merely a matter of happenstance: More specifically still, that the fundamental difference between human healthcare professionals and the AIs that they use is essentially a matter of history, because the technology of AIs has not yet been sufficiently developed. Therefore it is arguable that once this technical development crosses a certain threshold of complexity, AIs will also have minds, will be capable of conscious cognitive awareness and volition, and therefore will be persons.

The crucial question, therefore, is whether it can be shown that there is a fundamental difference between human persons and AIs, and that this is not a matter of technical complexity but of logic.

[23] Donceel JF, S.J. (1970). Immediate Animation and Delayed Hominization, *Theological Studies* 31: 1, 3–9.

[24] Schaffer J. (2014) What Not to Multiply Without Necessity. *Australasian Journal of Philosophy* 93 (4): 644–664.

Preliminary Considerations

And, indeed, there is an argument to this effect. It involves an appeal to mathematical and logical considerations. It goes as follows: To understand the logical nature of something is to model or to map that something in the understanding,[25] where the term "understanding" is merely a synonym for the cognitive parameters of the mind. If the entity in question is a finite material entity then, given its finite nature, its understanding or mind will have a finite structure or nature. Therefore, if an infinite logical structure is called an infinite set and a finite logical structure is called a finite set, then in order for a particular entity to be able to model an infinitely complicated logical structure it would have to be logically possible for an infinite set to be modeled in a finite set. That, however, is logically impossible because that would require that one and the same set could have conflicting logical structures. In other words, it would involve a contradiction.

However—so the argument continues—the set of numbers is transfinitely infinite, as is the system of relations that hold between them.[26] Therefore a finite material being cannot formulate and understand this logic and these relations: It would require that it be possible to model this logic and these relations in the material object; and, as was said, that is logically impossible. Consequently, it follows that if human beings can formulate and understand the logical nature of numbers and of the relationships that hold between them, then their minds cannot be material entities. More specifically, they cannot be identical with their brains. Consequently, while, as human beings, they have bodies and brains, their brains have to be what metaphorically may be described as the intersection points with the entities that do the actual understanding. These entities, in turn, are native to a different dimensional manifold in which the limiting parameters that characterise material entities do not apply.

Cantor, Turing, Gödel and Dualistic Metaphysics

The reason that these considerations are relevant to subject at hand and to the question whether there is a fundamental difference between healthcare professionals and the AIs that they use is that in 1878 George Cantor formulated what later came to be called the continuum hypothesis: i.e., that it is impossible to assign exactly one real number to each natural number in such a way that every real number gets thus

[25] *Cambridge English Dictionary*: "to model something" (mathematically)'

[26] Dedekind R. (1888), *Was sind und was sollen die Zahlen?,* Cambridge Library Collection (Paperback ed.), Cambridge University Press.

assigned.[27] Cantor's analysis is generally accepted as providing a mathematical account of the infinite logical structure of numbers and of their relations.

Furthermore, in 1950 Alan Turing established that entities in a three-dimensional space or manifold follow the laws of that manifold or space in the manner of their operation, and that these laws would prevent such three-dimensional entities from ever completing a processes that involved an infinite number of steps in a finite amount of time.[28] To put this differently, objects in the three-dimensional manifold or space can function either as chaotic systems, as nonchaotic systems, or in a combination of both. However, no matter how they function, their operation will be subject to the laws of that manifold. Material entities therefore, whether mechanical or biological in nature, will be subject to Turing's analysis. Consequently, for a material entity to capture the logic of numbers and of the relationships that hold between them, and in order for a material entity to be able to express that logic in arithmetic terms involving natural and real numbers, it would have to engage in an infinite number of steps in a finite amount of time. And, as Turing proved, that is logically impossible.

However, in 1931 Kurt Gödel, in his *On the Consistency of the Continuum Hypothesis and Formally Undecidable Propositions of Principia Mathematica,*[29] showed that no consistent system of axioms whose theorems can be determined on the basis of an effective procedure is capable of proving all truths about the arithmetic of natural numbers: In other words, there will always be statements about natural numbers that are known to be true but that are unprovable. Further, he showed that while his proof was valid, the system by which he demonstrated this could not demonstrate its own consistency. Gödel, of course, understood his proofs, as did many mathematicians and logicians after him.

Arguably, then, when this fact and the results of the preceding reasoning are applied to the issue of brains, they entail that human cognitive awareness cannot be explained in purely three-dimensional terms. For, brains, being three-dimensional objects, are subject to the laws of nature which, given Turing's analysis, means that they could not capture in a finite number of steps the mathematical logic that structures the nature of numbers and of the relationships that hold between them. Nor, even if they could symbolize these calculations, could they use understand that logic or use it in cognitively meaningful calculations. In other words, just like a laser printer, they could produce symbols that could be interpreted as being cognitively significant and as having these implications. However, just like the printer, they could not understand them. It therefore follows that human minds cannot be identical with human brains.[30]

[27] Cantor, Georg (1878). Ein Beitrag zur Mannigfaltigkeitslehre. *Journal für die Reine und Angewandte Mathematik.* 1878 (84): 242–258.

[28] Turing A. (1950). "Computing machinery and intelligence" *Mind* LIX (236): 433–460.

[29] Gödel K. Über formal unentscheidbare Sätze der Principia Mathematica und verwandter Systeme I., *Monatshefte f ur Mathematik und Physik* 38 (1931), 173–198.

[30] For a similar analysis, see Penrose R. (1994). *Shadows of the Mind.* Oxford University Press.

A possible counterargument to this could be based on the very fact that Gödel developed his theorems. After all, so it could be argued, Gödel used his brain to develop his proof. Human brains, however—Gödel's included—are material entities. Therefore even though the human brain is exceedingly complicated, it has to function in keeping with the laws of nature. This, in turn, means that it has to function essentially like a Turing machine. Consequently, the claim that the existence of Gödel's proof establishes that human minds cannot be identical with human brains is mistaken.[31] Accordingly, it is unnecessary to postulate a non-material metaphysical realm to which the human mind belongs.

Arguably, therefore—and applying these considerations to the question whether there is a fundamental difference between human minds and AIs—to deny that AIs could in principle have cognitive awareness is unjustified. Once they become sufficiently complicated in their structure, AIs would have awareness as well. Consequently, when the AIs that are used in healthcare reach this level of complexity, they then would become agents in the full sense of that term and not merely be tools, and they and their actions would also be subject to ethical evaluation.

Analysis

However, this counterargument may not be quite as telling as at first glance it may seem. For, to argue thus is to confuse what something is as a material entity with what it means. However, as Alfred Korzybski stated 1933, "a map is not the territory it represents"[32]; and as Wittgenstein already pointed out in 1921 in his *Tractatus Logico-Philosophicus,* an expression is not its meaning. It is merely a sign.[33] And as he elaborated further in his *Philosophical Investigations,* an expression, symbol or sign can be interpreted in many different ways but it does not carry its meaning in itself. It only has meaning if someone gives meaning to it.[34] Therefore while an algorithmic process may give rise to the totality of signs, symbols or expressions in which Gödel's proof is formulated, that is not the same thing as deriving the logical meaning that is expressed by the signs, symbols or expressions. To paraphrase Korzybski and Wittgenstein, it is to confuse the map with the territory: the meaning of an expression with the expression itself.[35] Consequently, the apparent refutation of the dualistic metaphysical hypothesis that was suggested above turns out to be unsuccessful.

---

[31] Landau, L.J. (1998). Penrose's Philosophical Error. In: Landau, L.J., Taylor, J.G. (eds) *Concepts for Neural Networks. Perspectives in Neural Computing*. Springer, London.

[32] Korzybski A. (1933). *Science and Sanity*, p. 58.

[33] Wittgenstein L. (1921). *Tractatus Logico-Philosophicus*, 3.31 ff.

[34] Wittgenstein L. (1953). *Philosophical Investigations* 4554 *et pass.*

[35] *Ibid.*, 38 ff.

When these considerations are applied to the issue at hand, however, it immediately follows that if the functioning of an AI was isomorphic with the functioning of the brain of a human being—and more particularly to the matter under discussion, if the functioning of an AI was isomorphic with the functioning of the brain of someone who understood Gödel's proof—then it would seem to licence the conclusion that if a dualistic metaphysical entity that was non-material in nature interfaced with the material body of a human person, then this would also be the case with an AI that was functionally isomorphic with the brain of such a human being. Also, and with due alteration of detail, it would licence the conclusion that if this functional-equivalence condition was met by a chip-replaced human being, then the individual would still be a person, the fact that their neurons had been replaced by chips notwithstanding.

An Issue and its Resolution

At this juncture, however, there arises another issue. More specifically, the analysis that was just criticised was based on the fact that Gödel et al. understood Gödel's proof. This supposedly shows that their awareness was not merely a brain-function but evidence of non-material metaphysical entities that, so to speak, merely used their material bodies to produce or deal with the symbols that expressed the proof. However, if the counterargument that was just adduced was valid, then it would seem to entail that human beings in general would have to be able to understand—or even be able to reproduce in a cognitively conscious manner—Gödel's proofs in order to satisfy the functionality criterion and to count as persons in this dualistic sense. And the reason that this seems to be problematic is that very few individuals who are not trained in logic and higher mathematics even understand the theorems, let alone are able to develop them. In fact, this holds true of most human beings. Therefore, since most human beings would not be able to do so, it would seem to entail that most human beings would not be persons in the full-blooded sense of that term, that they would merely be Turing machines, and that therefore their minds would not be non-material metaphysical entities of the sort that the preceding argument suggests. Consequently the conclusion that ultimately there would be no difference between most human beings—inclusive of most healthcare professionals—and AIs would stand after all.

An obvious reply to this last critique, however, is that it is not probative either because it rests on the assumption that individuals who do not understand something lack the capacity to do so. That assumption, however, is mistaken. The failure to understand something may simply be the result of insufficient training, or it may simply be the result of not having had the opportunity to become familiar with the relevant concepts—or, finally, of simply not being interested in them. Therefore in order for this critique to be valid, it would have to be shown on independent grounds that the failure on the part of most human beings to understand Gödel's proof was indicative of a lack of capacity to understand it. And this, in turn, would require that it would have to be shown in general terms that the failure to do something entails the inability to do the matter in question. In fact, however, this is not true. So, for instance, an individual's failure to actually run a marathon does not entail that they lack the ability or capacity to run a marathon, any more than their failure to speak Mandarin entails that they lack the capacity to speak Mandarin. In other words, not doing something does not in itself establish an absence of the capacity to do it. That would have to be shown on independent grounds.

When these considerations are applied to the issue at hand, they entail that in order for the preceding critique to be effective it would first have to establish that the failure on the part of most human beings to understand Gödel's proof was the result of their brains being neurologically insufficiently complicated, either in structure or in functioning, to be able to reach such an understanding. That, however, would require an in-depth and complete analysis of the brains both of the individuals who currently understand Gödel's proof and of the individuals who currently do not understand it, and showing that they are structurally and functionally distinct in the relevant areas. In the absence of this, at best the critique can establish the conclusion that this failure to understand *may be* indicative of a lack of capacity. That, however, does not amount to proof that it is the result of a lack of capacity. Consequently, as a counterargument to the initial analysis it fails.

A General Critique

At the same time, however (and as was noted before) the initial argument itself does not establish that human beings and AIs are fundamentally distinct and that, unlike the latter, the former are dualistic metaphysical entities. At best, it establishes that in some cases that *may* be the case. Therefore when all is said and done, it seems that the analysis as it has progressed so far, both on the pro and the con side, is inconclusive with respect to any fundamental difference between human beings such as healthcare professionals and the AIs that they use. To repeat, the analysis as it has progressed so far merely establishes that it may be possible that this is the case. However, as was said, the mere possibility that something is the case does not establish that it is the case, nor is it sufficient to base a policy on it, let alone an ethics policy that ultimately sets the framework for how health care that uses AIs is delivered.

The very inconclusiveness of the preceding analysis therefore raises the question whether the approach that has been adopted so far—an approach that involves talk about logical modeling, sets, subsets, Turing machines and Gödel's proofs, etc.—is misguided, and that adopting a different approach might lead to a clear answer to the general question whether there is a fundamental distinction between human and AIs, and to the particular question whether human healthcare professionals are persons whereas the AIs that they use are merely tools that they employ.

A Linguistic Approach

And indeed there is an alternative approach for dealing with the issue: It is to argue that the question whether someone is a person is simply a matter of language—of how the word "person" is used—and that logic, metaphysics and ethics really do not enter into the picture. In other words, one might simply follow the analysis that was proposed by P.F. Strawson when he said that "the concept of a person is the concept of a type of entity such that both predicates ascribing states of consciousness and predicates ascribing corporeal characteristics, a physical situation &c. are equally applicable to a single individual of that type."[36]

On this approach, human beings and AIs—and consequently healthcare professionals and the AIs that they use in their practice—are fundamentally distinct types of entities simply because the latter are legitimately called machines and only the former are legitimately called persons. Moreover, this approach would also licence the conclusion that once an individual was identified as a person then, no matter how they were altered or modified, they would remain a person until they are dead. Therefore it would also deal with the issue of what to say in the case of chip-modified human beings: If initially they were legitimately called persons, then they would always be persons, just like someone whose heart was replaced by a mechanical device would still be a person. The only exception would be if all of their cells were replaced by mechanical devices. Then they would be machines.

However, as attractive as this approach may seem, it would not be successful either, because it would merely amount to an evasion of the issue and to a begging of the question. For, the core question—whether healthcare professionals are fundamentally different from the AIs that they use—is not answered by suggesting that, as a matter of linguistic convention, a person is an entity to which it is legitimate to ascribe consciousness, volition as well as corporeal characteristics, whereas this is not the case with AIs. What is needed is an indication and justification of the criteria on the basis of which such predicates can be applied. And here, linguistic habits or standards do not provide an answer, for linguistic habits and standards are matters of convention—which leaves it entirely open whether such a convention reflects the nature of reality. So, for instance, there is a series of linguistic conventions that determine whether talk about warlocks and witches having magical powers is used appropriately, but this does not entail that in fact there are entities like witches and warlocks or that they have magical powers. Likewise, there is a series of conventions that govern the use of the term "jiva", and there is a whole framework within which appropriate jiva-talk is embedded, but this does not entail that the concept of jiva itself is valid.[37]

---

[36] Strawson PF. (1959). *Persons*. Routledge; pp. 101 f.

[37] Paramtattvadas S. (2017). *An Introduction to Swaminarayan Hindu Theology*. Cambridge University Press.

An Alternative Approach

At the same time, to suggest that the initial approach should be entirely abandoned would also be premature. Abstracting from the issue of indeterminacy, the logic that was involved in this approach was based on the assumption that the criterion in question—namely the capacity to produce and to understand something like Gödel's proofs—was a necessary condition that must be met in order for an entity to be considered a person and to ascribe a dualistic metaphysical nature to it. However, it is arguable that this is a mistake because it draws the frame of the analysis too narrowly. While being able to reproduce and understand something like Gödel's proof may certainly be indicative of having the capacity for cognitive awareness, it would not identify a necessary condition for being cognitively aware, and it would also not address the question of what cognitive awareness itself amounts to.

In other words, it is arguable that the mistake in the preceding approach does not lie in treating the capacity to produce or understand Gödel's proof as a *sufficient* condition but in treating it as a *necessary* condition. At most, that would indicate where, in the realm of persons, they would rank relative to intellectual capacity. And if that is correct, then there would be no reason to deny personhood to human beings such as healthcare professionals, most of whom neither understand nor can produce the relevant proofs. And, with due alteration of detail, the same would apply in the case of the AIs that are used by healthcare professionals.

To unpack this train of reasoning, it is necessary to go back to the question of what is meant by saying that an entity is aware, and whether that is different from saying that it is consciously cognitively aware. To be sure, the notions are related. Thus, an entity that was not consciously aware would not be cognitively aware except in a metaphorical sense. However—picking up on an issue that was mentioned before[38]—it is arguable that in order for an entity to be consciously cognitively aware is, among other things, to be self-aware: In other words, that only an entity that is self-aware can be said to be cognitively aware, regardless of what its level of cognitive awareness may be.

The Issue of Self-Awareness

But this immediately raises a whole series of other questions: Why should one accept this as a necessary criterion in the first place? Also, as was asked before, must the entity to which personhood is ascribed always be self-aware in order to count as a person? And, quite aside from all of this, it raises the question of what self-awareness itself amounts to, and of how the presence of self-awareness could be determined.

Part of the answer to these questions could be that, purely logically speaking, self-awareness is the ability of an entity to model itself within itself, because

[38] P. 6, *supra*.

self-transparency is characteristic of self-awareness.[39] Moreover, it is the ability to do so in a finite amount of time and without involving a vicious infinite regress.[40] That ability, however—so the argument—transcends the capacity of AIs as they exist at the present time and as they are used by healthcare professionals. Consequently, if and when the AIs that are used in the delivery of healthcare reach this level of complexity—for instance, if this was the case with AIs like ChatGTPs that are currently used in some forms of eHealth—then they would also count as persons and they would be more than mere tools. Therefore ethical considerations would then apply to them and to their actions as well.

Once again, however, this sort of answer only amounts to a mere restatement of the issue. It simply stipulates that the capacity for self-awareness must be present in order for an entity to count as a person. However, it still does not indicate how one could tell whether this was the case. Nor does it show how a claim of self-awareness was grounded. And inviting the entity in question to respond in positive or negative terms upon being asked whether they are self-aware would not settle the matter either. For, a ChatGPT, even as it is currently constructed, could do so. Moreover, the truthfulness of a positive response is not guaranteed by the validity of the response (as neither is the truthfulness of a negative response). Nor would it conclusively indicate that the answer was not a programmed answer.

Volition and Values

It is arguable, however, that the framework of the whole preceding discussion has still been far too restrictive. For, as was suggested in Chap. 1, a person in the full-blooded sense of that term is an entity that has the capacity for both conscious cognitive awareness and volition.[41] Therefore while a purely material entity might have the capacity for cognitive awareness in this full-blooded sense of that term, if it lacked the capacity for volition it would not count as a person. Therefore, if an AI did not have the capacity for volition—and this includes AIs such as those that currently perform surgical interventions or that diagnose health conditions once they are engaged[42]—it would not count be a person, the complexity of its actions notwithstanding.

---

[39] Werning M. (2010). Descartes discarded? Introspective self-awareness and the problems of transparency and compositionality. *Consciousness and Cognition* 19:3, 751–76; Byrne A. (2018). *Transparency and Self-Knowledge*, Oxford University Press See also Sartre JP. (1936). *The Transcendence of he Ego, passim.*

[40] Cameron, Ross (2018). "Infinite Regress Arguments". *The Stanford Encyclopedia of Philosophy*. Metaphysics Research Lab, Stanford University.

[41] Cf. Chap. 1, pp., 9 ff.

[42] Hamet P, Tremblay J. (2017). Artificial intelligence in medicine. *Metabolism*. 69S:S36-S40; Itchhaporia D. (2022). Artificial intelligence in cardiology. Trends *Cardiovasc Med*.;32(1):34–41.

If this is accepted, then the question becomes what the capacity for volition amounts to, and it entrains the further question why having that capacity would distinguish a purely material entity from a person. Moreover, it might also be asked how one could determine whether that capacity was present. Therefore, it would also involve the question of evidence.

The analysis that was presented in Chap. 1 and above seems to provide an answer to the first of these questions: The capacity for volition is the ability to choose among available options. However, as was also pointed out in Chap. 1 as well as above, choosing as opposed to responding in a more or less sophisticated stimulus-response manner to a given situation involves deciding to do so freely and on the basis of one's values. This, in turn, not only means that the relevant entity must have values but also that it must have free will. After all, to choose means "to select freely."[43] Consequently, if the entity is constrained in this regard, then its choice will not be a choice in the actual sense of that term. Therefore, this lack of constraint—this freedom to select—structures the very framework of decision-making of an entity that has free will: In other words, it then has volition as that term is properly understood. To be sure, this does not mean that such an entity *must* select among the options that present themselves, or that it *must* act: It may decide not to choose or not to act at all, even if it subscribes to values that would in principle identify the direction of their choice.

Values: Some Relevant Considerations

Once more, however, this answer merely complicates the issue. By stating that an entity has free will only if has the capacity for choosing independently of being constrained to automatically do so by the laws of nature or by its design merely transfers the issue to the question of what role the possession of values plays in this regard. Moreover, it raises the question what in fact values are: Why should values be considered to be distinct from factors that result in law-like and deterministic actions?

In order to answer these questions, it may be useful to once more return to Chap. 1. There, it was suggested that insofar as persons are concerned, values should be understood as action potentials.[44] Not, however in the sense of action potential as this term is understood in the physical sciences but as providing a non-determining inclination to act in a certain manner. In other words, it was then suggested that values function as action potentials in the sense of providing an oriented motivational impetus for engaging in a particular action on part of the individual who holds these values, but without thereby constraining that individual to act in a particular manner.

---

[43] *Merriam-Webster Dictionary*, choose" (definition).

[44] Cf. Chap. 1, p. 10 f.

When values are understood in this sense, they are fundamentally different from action potentials as these are understood in the purely material world and in the physical sciences. In the latter, an action potential is the energy that is held by an object relative to other objects.[45] Thus, an electric potential is the difference in electric pressure or electric tension between two distinct entities, and it is defined as the amount of work that is needed per unit of charge to move a test charge between the two.[46] This may be expressed in numerical terms. More importantly, however, is the fact that whether and how such a potential is actualised is solely a function of its strength in relation to the potential that is inherent in other entity to which it stands in relation.

Some Differences

An important difference between a purely material entity and a programme that is written into it on the one hand and a person on the other, therefore, is the fact that a purely physical entity or a programme that is written into it cannot decide not to actualise its action potential once the relevant situation presents itself. In other words, the actualisation of its action potential is not a matter of choice. On the other hand, when values are understood as action potentials in the case of persons, no such determination exists. Whether the person who subscribes to a given value acts on it is a matter of choice, and it lies entirely under their control whether they will act in accordance with that value or, indeed, whether they will act at all.

A further difference between the two sorts of cases is that if there is no difference between two purely material entities in the action potential that exists between them, then the purely material entities will not act. The case of electric potentials that was mentioned above once more illustrates this. By contrast, a person can decide to act even when the situation that presents itself involves no discernible difference between the options that obtain. In fact, a person can decide to do something precisely because the situation in which they find themselves is undifferentiated in any material, valuational or cognitive sense: For instance, because they find it boring and they want to "add spice" to their life. As was said, nothing like this is possible for purely material entities.

In reply, it could be argued an AI can be programmed in a way that, when confronted with what might be called a situation of equilibrium, it will act by randomly picking one of the possible options that exist at that point in time. In that case it would engage in an action after all, thus paralleling the behaviour of the "bored" person. As well, a person, when faced with a situation of equilibrium may simply be unable to act and be, so to speak, perplexed. Arguably, therefore, it would not seem to be justified to take these sorts of situations as evidence of a fundamental

[45] Jain, Mahesh C. (2009). "Fundamental forces and laws: a brief review". *Textbook of Engineering Physics*, Part 1.

[46] *Encyclopedia Britannica*, "electric potential".

difference between purely material AIs and human beings, and that the latter's actions required a dualistic metaphysical explanation.

However, there are two aspects of a person's valuational framework that find no parallel in an AI and that do seem to require a dualistic metaphysical explanation for the nature of human persons after all. The first aspect is that the values or action potentials that belong to person are not inherent in that person in the sense that they are not a function of their very nature or constitution. The second aspect is that persons can subscribe to or be motivated by contradictory action potentials or values.

The Noninherence of Values

To expand on the first: If an entity is purely material in nature then, whether an action potential obtains in it as well as the nature and strength of that action potential is determined by how it is constituted. That is to say, it is functionally related to its material properties or characteristics. Therefore when it changes in its material nature, so does its action potential. Therefore two purely material entities such as two AIs that are used in healthcare and that are constituted in the same way, when confronted with identical situations will have the same action potentials and therefore will react in the same way to the stimuli to which they are subjected. On the other hand, when their constitution changes, then their action potentials will change as well because they are functionally determined by their constitution or nature. Moreover, the changes in their action potentials will be law-like and will in principle be predictable.

However, that is not the case with respect to human persons. What values human beings have is not functionally determined by their physical constitution. Human persons who fundamentally differ in their physical or biological constitution may subscribe to the same values despite their material differences. This is so because the values to which they subscribe to not depend on nor are predictable on the basis of their material natures, qualities or characteristics. Therefore, even though distinct persons may differ fundamentally with respect to their sexual or other physical characteristics, they may share the same values. Moreover, even though they may go through a process of becoming sexually mature or senescent, they may continue to subscribe to the same values despite the fact that their physical characteristics change.[47]

Secondly, a person can change the values to which they subscribe even though nothing has really changed in their physical make-up. Thus, someone who previously subscribed to the Principle of Utility and therefore to a utilitarian value framework may change their mind and decide to make their decisions and choices on the basis of the values that are central to Judaic, Islamic or Hindu beliefs without

[47] Bayly, B. L., & Bumpus, M. F. (2020). Patterns and Implications of Values Similarity, Accuracy, and Relationship Closeness between Emerging Adults and Mothers. *Journal of moral education*, 49(4), 496–511. https://doi.org/10.1080/03057240.2019.1669545

changing their physiology. This is sometimes referred to as "conversion".[48] Likewise, a person may decide to abandon a scientific world-view and accept a magical world view, where these world views involve fundamentally distinct and logically incompatible ways of understanding so-called causal relations and the nature of reality, all without undergoing a physiological alteration in their nature. They will therefore react in a new manner to the situations in which they find themselves despite the fact that their biological makeup has not changed.

The Implications of Logic

Furthermore—and to expand on the second reason that was mentioned above—a purely material entity is subject to the laws of nature. Not, however, merely with respect to how it reacts to the stimuli to which it is subjected but also in the sense that the logic of its very nature has to be consistent with the logic that structures the nature of reality itself. And that logic cannot be contradictory because, as already Aristotle said[49] and as Bertrand Russell reiterated,[50] contradictory entities—and that includes reality itself—cannot exist.

This, of course, does not mean that a purely material entity cannot produce or engender something whose significance is incompatible with the facts of reality. Thus, a typewriter may be used to produce sentences that are grammatically correct but that are logically incoherent. Likewise, a computer programme like a generative pre-trained transformer (GPT)[51] can also produce output whose significance is incompatible with the facts of reality. However, the process by which these are produced must be logically coherent, and must be in keeping with the laws of nature.

Still, these last considerations do not negate the fundamental difference between AIs and human persons. That is to say, an AI, being a computer programme that is performed by a material entity, cannot violate the laws of logic. That does not, however, mean that it an AI cannot consist of subroutines that function according to different logics. For instance, one could design an AI in such a way that one subroutine calculates on the basis of modal logic, another subroutine on the basis of fuzzy logic, and still another by using deductive logic, etc. However, it means that such an AI cannot have a single operating system that simultaneously functions in these logically distinct ways. Instead, what is then called the AI would actually be an overarching switching programme that switches from one subroutine to another: much like a car that is switched by its operator from going forward to going

---

[48] Stark, Rodney and Roger Finke. "Acts of Faith: Explaining the Human Side of Religion." University of California Press, 2000. p.114.

[49] Aristotle, *Metaphysics* 1055b2.

[50] "The Philosophy of Logical Atomism," reprinted in *The Collected Papers of Bertrand Russell*, 1914–19, Vol 8., p. 228.

[51] Jovanović M and Campbell M. (2022). Generative Artificial Intelligence: Trends and Prospect, *Computer* 55:10, 107–112.

backwards and vice versa. Therefore, to say that such an AI would in fact be operating with mutually incompatible logics would be to confuse the functioning of the distinct subroutines with the functioning of the overarching AI, and to say that they are one and the same system would be to claim that they constitute one system because they exist in one and the same computer.

Human persons, on the other hands, can simultaneously think in contradictory ways. For instance, they can love as well as hate another person at one and the same time. While different cerebral circuits may be involved in these distinct states, the awareness of the person who has both attitudes is simultaneous: Otherwise they would not be aware that they, as an individual, both love and hate the other person. Another example is the case of so-called ambiguous figures that are bistable. While different neural brain circuits may be involved in these distinct perceptions,[52] the awareness or consciousness of the perceptions is contained in one and the same consciousness since otherwise there could not be a unified awareness of the fact that they are ambiguous.

In opposition to this it might be argued that it is premature to conclude that therefore AIs, being the operating systems of material objects, cannot function according to different logics. While this may hold true of computer systems that are programmed into standard computers, this need not be true of AIs that are programmed into computers that are made up of qubits. More particularly, standard computers are made up of standard electronic chips, and these are binary in their functioning: I.e., they can only be in on or an off state at the same time, but not in both states simultaneously. By contrast, qubits are not subject to this limitation.[53] Therefore an AI that was programmed into a computer that was composed of qubits could be in distinct states at one and the same time. Consequently it could be aware of conflicting states and therefore have an ambiguous awareness, just like a human person.

The tenability of this reasoning will be examined more specifically in Chap. 6, which deals with artificial general intelligences in healthcare. For now, suffice it to say that to argue thus is to confuse a material state with an awareness. In other words, while each material basis of such an AI—i.e., its qbits— would be in the distinct states simultaneously, these states themselves would be logically distinct, and neither one would include the other. However, in order to be a state of simultaneous awareness, the distinct states themselves would have to be unified in a single state, and that is logically impossible.

---

[52] Rodríguez-Martínez GA, Castillo-Parra H. (2018). Bistable perception: neural bases and usefulness in psychological research. *Int J Psychol Res (Medellin)*. 11(2):63–76.

[53] B. Schumacher (1995). "Quantum coding". *Physical Review* A. 51 (4): 2738–2747.

Conclusion

The present chapter has outlined more specifically what is meant by the term "healthcare professional" and why, in the rest of this study, it will be understood in this way. It was also suggested that while all healthcare professionals are subject to ethical considerations, what these considerations are and how they apply how differs on the basis of how they are engaged in the profession of healthcare.

A major part of the reasoning in this chapter was devoted to indicating how and why healthcare professionals differ from the AIs that they use, and it also expanded on the reasoning that underlay the claim that was made in Chap. 1: namely, that while ethical considerations directly apply to healthcare professionals, they apply to the AIs that they use only in a derivative sense.

At the present time, there only exist expert systems or AI-2 s, not AI-1 s. These expert systems are tools, not agents in their own right: not even when they are extremely complicated and when they function on their own once they have been activated. Whether there will ever be AI-1 s is another issue. The preceding analysis has indicated what would be necessary for that to be the case. The final chapter will explore this further. The next chapters, however, will proceed on the assumption that as they are currently constituted, AIs are what were initially called AI-2 s, and that they are mere tools. The subsequent discussions will therefore outline and explore the ethical considerations that are relevant in the development and use of such AIs by the various parties who are engaged in the delivery and planning of healthcare.

Chapter 3
AIs and Hands-on Healthcare Professionals

Abstract The present chapter considers the ethical implications of the design and use of AIs by hands-on healthcare professionals. Since AIs are informatic tools whose processes rely on data, it begins by outlining the notion of data-needs as this applies to hands-on healthcare professionals and the implications that this has for their use of AIs. It then indicates how this can be integrated into the ethically appropriate design and functioning of the AIs that they use, including so-called deep-learning AIs.

Introduction

As was indicated in the previous chapters, AIs as they currently exist are electronic informatic tools that are used by modern hands-on healthcare professionals as well as by modern healthcare administrators and policy makers in carrying out of their tasks. While the roles of these professionals overlap to some degree, they nevertheless are distinct in focus and involve some different factors. These differences impose different technical requirements on the AIs that they respectively use. Moreover, the distinct roles in which these different types of professionals engage also have distinct ethical implications, which also impose different requirements on the structure and use of their respective AIs.

Thus, as a matter of general principle, hands-on healthcare professionals stand in a fiduciary relationship to their patients. However, what this fiduciary relationship ethically entails for hands-on healthcare professionals who are in direct and personal contact with their patients is slightly different from what it entails when their interaction is with patients who are geographically separated from them and whose interaction is electronically mediated. It is even somewhat different when the professionals act as consults for other healthcare professionals who are directly dealing with specific patients, for then it is a physician-patient relationship at one remove. All of these variations impose different requirements on the design and functioning of the AIs that they use.

E.-H. W. Kluge, *The Ethics of Artificial Intelligence in Healthcare: From Hands-on Practice to Policy-making*, The International Library of Ethics, Law and Technology 25,
https://doi.org/10.1007/978-3-031-72837-2_3

With due alteration of detail, similar considerations apply to healthcare administrators and policy makers. For instance, the ethical issues that arise with respect to the AIs that a healthcare administrator who is responsible for the administration of a unit such as a hospital or a clinic are somewhat different from those that arise for an administrator who is responsible for a health district that spans a larger geographic area that has a diversified demographic makeup. Likewise, the ethical issues that arise with respect to the design of the AIs that are used by policy makers who are situated in a right-oriented socio-political setting are somewhat different from those that arise for policy makers who are situated in a socio-political setting that views healthcare as a commodity. To say nothing about the ethical implications for the design of AIs that are used by healthcare providers such as *Médecins Sans Frontières*, or about the implications that any of this has for the health informatics professionals who ultimately are responsible for the technical aspects of the AIs that are used in all of these distinct ways.[1]

This chapter, however, will only address some of the more important ethical issues that surround the design and functioning of AIs that are used by hands-on healthcare professionals. What is involved in the case for healthcare administrators and policy makers will be the subject of the next chapters.

A Few Distinctions

Human vs. Non-human Healthcare

Before actually dealing with these issues, it may be appropriate to once more draw a few distinctions. One of the more important distinctions concerns hands-on healthcare professionals who deal with human beings and hands-on healthcare professionals who deal with non-human animals. The reason why drawing this distinction is relevant is that, as was said in the preceding chapters, the concept of health applies to all biological beings and in particular to all animals. And while the technical requirements for the AIs that these healthcare professionals use are in many ways similar, the ethical issues that arise with respect to the design and use of their AIs are quite distinct.

For, if the analysis that was presented in Chaps. 1 and 2 is valid, then ethical considerations in the traditional sense of the term "ethics" arise only when dealing with persons. Therefore, since non-human animals are traditionally not considered to be persons, any ethical considerations that arise in such a context are ethical

[1] Kluge E-H W. "Electronic Health Records: Ethical Considerations Touching Health Informatics Professionals", in Hsueh, Wetter and Zhu, eds. *Personal Health Informatics: Patient Participation in Precision Health*, pp. 467–484, Springer, 2022.

considerations only in an extended sense of the term "ethics".[2,3] Furthermore, even when non-human animals are considered to be persons, they generally are not considered to be persons in quite the same sense.[4,5] As well, it is arguable that even if ethical considerations are taken to apply to non-human animals, then such considerations are not based on the same ethical principles or, if they are based in the same ethical principles, that they are not based on these principles in the same way.[6] Therefore, since the focus of the present study is the ethical implications of the design and usage of AIs in the context of healthcare that deals with human beings, what follows will proceed independently of any ethical considerations that might arise in the design and usage of AIs that are involved in the healthcare of non-human animals or in the context of a laboratory setting.

Different Types of Data Needs

However, as was also indicated in the preceding chapter, even when the term "healthcare professional" is understood in this specific sense, what it denotes actually divides into distinct rubrics; and while these overlap to some degree in the domain of professional-patient interaction, they have distinct requirements and ethical implications even in the hands-on setting.

So, for instance, the profession of nursing is different from the profession of medicine, as is that of pharmacy, dentistry, midwifery or psychiatry, etc.[7] And while all hands-on healthcare professionals require access to patient records, that access has a different focus even when the records pertain to the same patients simply because the data that are relevant for their respective professional activities are somewhat different, and in many ways this difference has distinct ethical implications with respect to the design and functioning of the AIs that they use in carrying out their professions.

So, for instance, the data-needs of a psychiatrist are different from those of a primary care physician. Thus, while the mental or psychiatric functioning of a patient may be functionally related to their metabolism and may ultimately be related to their genetic makeup, the details of this are generally not relevant for

---

[2] Beauchamp TL. "Introduction," in Tom L. Beauchamp and R.G. Frey. *The Oxford Handbook of Animal Ethics*. Oxford University Press, 2011.

[3] Loeb JM, Hendee WR, Smith SJet al. "Human vs Animal Rights: In Defense of Animal Research", JAMA. 1989;262(19):2716–2720.

[4] "The Moral Status of Animals," *Stanford Encyclopedia of Philosophy*, accessed 04/07/2022 at https://plato.stanford.edu/entries/moral-animal/

[5] Karcher, Karen. "The Great Ape Project" in Marc Bekoff (ed.). *The Encyclopedia of Animal Rights and Animal Welfare*. Greenwood, 2009.

[6] Russell, W.M.S. and Burch, R.L., (1959). *The Principles of Humane Experimental Technique*, Methuen, London.

[7] Trimble M. The Profession of Medicine and its Rivals. *Ulster Med J.* 2016 May;85(2):76–9.

psychiatry. On the other hand, such data are importantly relevant for the hands-on physician who diagnoses a patient and who considers what—if any—intervention to propose to the patient or to their duly empowered substitute decision-maker for their consent. This difference, therefore, ethically entails that the AIs that they respectively use in the conduct of their professions should be differently structured: Not merely with respect to the expert systems that they may use when diagnosing their patients but also with respect to the patients' data that are accessed by them and by their AIs.

Or, to use another example, the kind of care that nurses deliver is in many ways data dependent, just as is that of physicians. For, only if they know the condition of the patients with whom they are dealing, and only if they know the etiology of their patients' condition can they provide appropriate healthcare. This means that nurses who deliver hands-on patient care have to be able to access their patients' records. This type of access, however, is different from the type of access that is necessary for the physicians who are, so to speak, in charge of the patients and who oversee their treatment. Therefore, such differences are appropriately reflected in the design and functioning of the AIs that these distinct professionals use, so that they can perform their profession without worrying about their distinct ethical implications.

However, since AIs are electronic systems that do not interact with patients in a phenomenal sense but on the basis of electronically encoded data, it may be appropriate to expand on why and how this is the case by considering first how, informatically, a patient presents in the healthcare setting: That is to say, by considering first how a patient presents to a healthcare professional from the perspective of data and relative to the nature of the professional-patient interaction, and then translating this into terms that apply to the AIs that are used by such professionals.

The Patient as Data-Space

Purely informatically speaking, a patient presents to a healthcare professional as a totality of possibilities of distinction—what may be called a data-space—that consists of the qualities or characteristics of that particular patient. In that sense—and again speaking only informatically—every patient presents as a unique data-space because every patient is unique as an individual and is distinct from ever other person.

Each of a patient's data, in turn, is embedded in a logical sub-space in virtue of its being a datum of a specific type. Thus, a blood pressure datum is embedded in the sub-space of cardiac function, whereas urea and creatinine data are embedded in the sub-space of renal function, etc. In other words, these sub-paces deal with what might be called the organ- and behaviour-relative characteristics of the patient. These, in turn, are logically embedded in ever more general data-spaces until the total data-space of the patient is reached, where this contains all of the specific patient-relative data sub-spaces. Ultimately, therefore, this data-space will include all the data sub-spaces that deal with the molecular, genetic, behavioural and morbidity conditions of the patient in question. Finally, informatically speaking, there will be as many of such sub-spaces as there are types of data, and all of these sub-spaces will be logically related to each other because they are sub-spaces that constitute the overall data-space that is the patient.

Of course this does not mean that the healthcare professional to whom a patient presents is aware of all of these data and of the relations that exist between them. What the professional is aware of is functionally determined by their individual insight, the instrumentation or means of observation that they use as well as by the nature of their training and by the investigative effort that they expend. All it means is that, considered from a purely informatic point of view, this is how a patient presents to the professional.

Consequently, not only will the same patient present as a different data-space to the same hands-on healthcare professional on different occasions—for instance, when the patient exhibits different qualities or characteristics—but they will also present as a different data-space to different healthcare professionals even when these professionals are of the same type; and they will present as an even more profoundly different data-space to healthcare professionals of different specialties. As well, these data-spaces will change for individual professionals as their own experience changes and as the professional's scientific knowledge, the instruments that they use and relevant research changes.

EPRs, AIs and Patient Profiles

All if this has implications for the design and functioning of AIs that are used in healthcare at the hands-on level. Thus, AIs, when they are used as diagnostic tools, will either be connected to the instruments that are used to generate data about the patient, will access the electronic patient record (EPR) that contains these data, or both.

An EPR, in turn, may be viewed either materially or informatically. Viewed materially, it is an electronic entity with certain characteristics; viewed informatically, it is the totality of data that have been generated about a given patient. In other words, it presents the patient-relative data-space in electronic form. Moreover—and again speaking informatically—the data-space that is encoded in an EPR is the basis of the patient profile that is generated about the patient when the data in the patient EPR are correlated. This, of course, means that different healthcare professionals using different AIs may well generate different patient profiles even when they are dealing with one and the same patient and are using one and the same EPR.

Patient Profile as Patient Analogue

And again speaking purely informatically, this profile is a patient analogue. To be sure, it is not a strict patient analogue. As was said before, a patient constitutes a totality of possibilities of distinction, and not all of these may be included in the data that constitute the EPR, nor may all of these data be correlated into a single whole. Nevertheless, the profile functions as patient analogue insofar as the informatic aspects of the professional-patient interaction is concerned. It functions as patient analogue because it is on this basis that decisions about the patient are made by the professional even when the patient is actually present. For, the material patient is not what constitutes decision-relative data. The material patient is merely an entity: a thing. It is the data about the material patient as assembled into the patient profile that is the informatic entity on which the professional's decisions are based.

In other words, the profile is a model of the patient and, speaking decision-theoretically, it substitutes for the patient. And if the patient profile involves the data that are contained in the patient's EPR—which is usually the case, because the profile is generally not simply based only on the direct and immediate investigation of the patient but takes into account the patient's different characteristics as these are identified on different occasions—then the EPR partly constitutes the informatic and epistemic foundation of the patient profiles that are developed by healthcare professionals when making their diagnoses and healthcare decisions about their patients.

EPRs and AIs

When these considerations are applied to a hands-on professional's use of an AI, the primary fact is that the AI, being an electronic programme, cannot interact with a patient in a hands-on fashion like a hands-on human healthcare professional. It is restricted to the use of electronic data about the patient. In other words, it functions as a tool that correlates the patient data that constitute the patient's data-space and, when it is involved in an instrumental intervention, by electronically directing the instrumentation on whether and how to proceed.

This means that in order to function as the professional's tool, the AI must also have access to those parts of a patient's EPR to which the professional would normally have access.[8] This, of course, has privacy implications because the health-related data that are contained in a patient's EPR are data that are unique to the patient and can be used to identify the latter. Ethically, this is expressed by saying that patients are surrounded by a sphere of privacy that may be breached only with the patient's consent.[9] Not only that, these data may have ownership implications.[10]

Leaving aside the matter of ownership and of the technical quality of the purely diagnostic tools that are used to gather the data that make up the data-space of the

[8] For exceptions to this, see *infra* relative to innovative treatment and professional data-spaces.

[9] Kluge E-H W. (2020). *The Electronic Health Record: Ethical Considerations*. Academic Press, Chap. 11.

[10] Bartoletti, I. (2019). "AI in Healthcare: Ethical and Privacy Challenges," in Riaño, D., Wilk, S., ten Teije, A. (eds) *Artificial Intelligence in Medicine*. AIME 2019. Springer, Cham. https://doi.org/10.1007/978-3-030-21642-9_2

EPR, also leaving aside the issue of how precisely and how accurately the particular instrumentation has generated the data of the patient data-space, and likewise leaving aside the issue of whether that instrumentation was employed at the independent discretion of an AI (a matter that will be addressed in Chap. 6 when AIs will be considered as independent agents), this means that if an AI is to mirror the actions that would otherwise be performed by the healthcare professional, then it must reproduce in electronic terms the diagnostic and prognostic activities in which a human healthcare professional engages when diagnosing and prognosticating. The human healthcare professional, when doing so, accesses the patient's data-space either directly by interacting with the patient or indirectly by accessing the patient profile in the EPR that is the patient analogue or both. Therefore if an AI is to mirror—or even to replace—such human behaviour, it also must have access to the entire relevant patient data domain to which the professional would normally have access.

Data-Need Domains

At this juncture, an ethical issue emerges. It is the issue of the data domains to which hands-on healthcare professionals have legitimate access when interacting in a professional capacity with their patients, and it is an issue that has implications for the design and use of AIs in the hands-on healthcare setting. The issue is that of privacy.

As was said before and all other things being equal—the nature of the exceptions will be addressed in a moment—a patient is ethically surrounded by a sphere of privacy that may not be breached without their consent. If the type of patient-relative data to which the healthcare professional has legitimate access when engaged in their profession is called the data-need domain of that professional then, ethically speaking, a healthcare professional's data-need domain is not necessarily coextensive with the whole data-space of the patient or that is contained in the patient's EPR. Since this has important implications for the design and functioning of the AIs that they use, it may be worth clarifying further how this is the case.

To begin with—and as is only too obvious—different kinds of healthcare professions differ in their focus on the health conditions and healthcare needs of the patients with whom they interact. This means that while the patient data that they require in order to make diagnoses, prognoses and treatment recommendations may overlap, they will not be coextensive. If, using the vocabulary that was introduced a while ago, the data that they need in order to do so is called a data-need domain, then this may be expressed by saying that the data-need domains of different kinds of hands-on healthcare professionals are characteristically different. Since, as was said, patients are surrounded by a sphere of privacy that generally may not be breached without their informed consent, this means that when a hands-on healthcare professional interacts with a patient, the person-relative investigations that the professional may perform is ethically limited to the data that are relevant to the professional's data-need domain. For the same reason, the questions that the professional may pose to the patient are likewise limited.

Consequently, when a healthcare professional uses a patient's EPR, this restricts the professional's access to the data-space that is presented by the EPR. Therefore if a professional attempts to transcend their typical data-need domain without the patient's consent when accessing the EPR, this would normally constitute a *prima facie* invasion of privacy and hence, on the basis of the Principle of Autonomy which underwrites the right of privacy, it would be unethical. Consequently, if a healthcare professional made such an attempt without the patient's consent, a properly constructed EPR would not necessarily prevent this since the professional may have legitimate reasons for doing so; however, it would alert a duly empowered informatic authority of this transgression. That authority would then take appropriate steps to deal with the situation. Theoretically, these steps would include requiring the professional to justify their action.

When all of this is translated into the design and use of an AI, it comes out as the requirement that normally an AI that is used by a hands-on healthcare professional would contain a provision that would require an indication by the patient that the latter considered access to their EPR to be acceptable. That consent may be a derivative of the consent that the patient gives to the professional-patient interaction or it may be explicit. In either case however, and ethically speaking, it would be necessary because it would satisfy the overall informatic consent requirement that would legitimate an AI's intrusion into the patient's sphere of privacy. Moreover, and unless otherwise specifically instructed, the AI that the professional might use would access only those patient data that would fall within the data-need domain of the professional who was using it. If it was instructed to do otherwise, a properly designed AI would automatically alert both its user and a duly empowered informatic authority to this effect, who would then take appropriate steps to deal with the situation.

At the same time, however, the AI's access to the patient data that constitute the healthcare professional's data-need domain may be insufficient for the AI, as expert system, to be able to frame a diagnosis, let alone to formulate either a prognosis or make a treatment recommendation. For, in order to be able to do so—and just like its professional user—it might have to have access to data that deal with different kinds of treatment modalities that are not standard for the particular profession of its user but that are standard when they are provided by members of different healthcare professions when the patients present to them with a similar issue. Moreover, in order to evaluate this properly, it might also need to access the outcomes of these modalities in the case of these other patients. This, once more, has distinct ethically relevant quality, privacy and security implications for the design of hands-on AIs.

"In the Patient's Best Interest"

The matter goes further: The fiduciary duty that hands-on healthcare professionals have towards their patients stipulates that, all other things being equal, they should always act in their patients' best interest. The notion of best interest, however, is defined not simply in terms of what is considered to be the best outcome from a purely professional perspective but what is considered to be the best outcome consistent with the (competently held) values of the patient in question. Ethically, this is grounded in the Principles of Autonomy, Beneficence and Nonmaleficence. Therefore this may alter the treatment recommendations that a healthcare professional considers.

Consequently, an AI that is used by a hands-on healthcare professional should also have a parameter that identifies the patient's values, and the AI should also use these values to assign what effectively are action potentials to the various treatment options that are in keeping with these values. In other words, in a properly constructed AI that allows its professional users to fulfil their fiduciary obligation to always act in the best interest of their patients, the AI would turn the patient's values into gradients that structure the sub-spaces of possible treatment options. From these the AI would then select a particular treatment option and present it either to the healthcare professional as a treatment option or to the patient for their consent, as the case may be.

Further—and still under the heading of best interest—a patient may previously have undergone the same treatment or some other treatment that had a particular outcome. A conscientious healthcare professional would take cognizance of this and adjust their treatment recommendations accordingly. As well, such a healthcare professional would attempt to be current with developments in their particular area and with the relevant treatment recommendations as these are promulgated by appropriate medical authorities.

A properly constructed AI, therefore, would match this behaviour when it is used to assess the healthcare needs of a patient. Consequently, it would have to be able to scan the data relative to the standard interventions as well as relative to new developments and treatment recommendations. And as past studies have shown, AIs not only have the potential to outperform human healthcare professionals in this regard and to come up with more accurate diagnoses and prognoses,[11] they also have the potential to alleviate the current labour shortage of qualified healthcare professionals.[12]

---

[11] Killock, D. AI outperforms radiologists in mammographic screening. *Nat Rev Clin Oncol* 17, 134 (2020); Oszwald A, Wasinger G, Pradere B, Shariat SF, Compérat EM. Artificial intelligence in prostate histopathology: where are we in 2021? *Curr Opin Urol.* 2021 Jul 1;31(4):430–435; Groot OQ, Bongers MER, Ogink PT, Senders JT, Karhade AV, Bramer JAM, Verlaan JJ, Schwab JH. Does Artificial Intelligence Outperform Natural Intelligence in Interpreting Musculoskeletal Radiological Studies? A Systematic Review. *Clin Orthop Relat Res.* 2020 Dec;478(12):2751–2764.

[12] Scheffler RM, Liu JX, Kinfu Y, Dal Poz MR. Forecasting the global shortage of physicians: an economic- and needs-based approach. *Bull World Health Organ.* 2008 Jul;86(7):516-523B; World

Privacy and Security Implications

However, the design and use of AIs in this regard does not come without ethical problems. Thus—and focusing for the moment only on such issues as privacy and security—these attach not only to the technical design of the AIs insofar as they are machines but also with respect to their usage.

These issues essentially arise at the same time and in the same manner: They arise from the fact that the AI would have to go outside of the data domain of the EPR and access the data bases that contain the relevant information. Moreover, in order not to do so randomly but to focus their enquiry relative to the patient's condition, they would have to identify that condition. However, as research has shown, given three or more data points that are contained in an EPR it is possible to identify the patient themselves even if the data are anonymised.[13,14,15,16] Therefore unless the enquiry is in terms of general characteristics—which, as was said, may not be useful for identifying an appropriate treatment option for a specific case—or the enquiry is encrypted, the patient's privacy may potentially be breached by anyone who is aware of the AI's enquiry. Therefore, unless the AI's search is restricted to data bases that are in-house, or unless the AI's actions are otherwise shielded from being observed by someone who is not entitled to do so, the AI's actions themselves will be observable by an unauthorised party and therefore will give rise to privacy concerns. Consequently the very fact of making an AI responsive to appropriate best-interest considerations has privacy implications with respect to its design and operation.

Moreover, since the AI's actions would be in reference to a particular patient's condition, the patient's EPR would have to be accessed by the AI. Consequently, this opens up the possibility that even though the EPR itself was constructed in such a way that entry into it was somehow restricted to duly empowered individuals or to the AIs that are used by such individuals, and unless their actions are somehow protected from external observation, an unauthorised observer would be able to intrude through the AI's search action into the patient's EPR. Consequently, such an observer would not only be able to breach privacy but also to erase or modify the data, possibly even without any notice or indication. Colloquially, this is sometimes

---

Health Organization, Health Wworkforce, accessed 09/07/2022 at https://www.who.int/health-topics/health-workforce#tab=tab_1; WHO, The World Health Report, accessed 09/07/2022 at http://www.who.int/whr/2006/en/

[13] Emam KE, Dankar FK, Vaillancourt R, Roffey T, Lysyk M. (2009). Evaluating the Risk of Re-identification of Patients from Hospital Prescription Records. *Can J Hosp Pharm.*62(4):307–19.

[14] Porter CC. (2008). De-identified Data and Third Party Data Mining: The Risk of Re-identification of Personal Information. *Washington Journal of Law, Technology & Arts* 5:1, 1–8.

[15] Richardson V, Milam S and Chrysler D. (2015). "Is Sharing De-Identified Data Legal? The State of Public Health Confidentiality Laws and Their Interplay with Statistical Disclosure Limitation Techniques". *The Journal of Law, Medicine & Ethics.* 43 (1_suppl): 83–86.11.

[16] Higgins TL. Reidentification of Protected Health Information: Can the Risk Be Quantified*? Crit Care Med. 2021 Jun* 1;49(6):1003–1006.

referred to as hacking. Therefore unless the AI was constructed in such a way that it was entirely disconnected from the patient's EPR when it queried a data base, or some technical means was employed to prevent the possibility of such an intrusion into the AI's own operating system, the very use of the AI, no matter how efficient and effective it may otherwise be, raises this sort of ethical concern as well.

"All Other Things Being Equal"

One of the points that were made in passing was that a hands-on healthcare professional's duty to always act in the best interest of their patient is conditioned by the clause "all other things being equal." Since this limitation also applies to the tools that healthcare professionals use and therefore also applies to their AIs, it may be appropriate to clarify the significance of this phrase with respect to the actions of hands-on healthcare professionals, and then to spell out its implication with respect to the construction, functioning and use of the relevant AIs.

The conditioning clause itself refers to the claim that a hands-on healthcare professional's duty to always act in the best interest of their patient not only includes the requirement of adjusting their actions in accordance with the patient's values but also that what the professional does should not result in harm. This has traditionally been expressed by the phrase "above all, do no harm (*primum non nocere*)."

This duty, in turn, is grounded not only in the Principles of Nonmaleficence and Beneficence but also in the Principles of Impossibility. All of these have implications both for the healthcare professional who interacts with the patient and for the AIs that they use.

Impossibility

Focusing first on the Principle of Impossibility—and ignoring the question whether a healthcare professional could reasonably be expected to be aware of all possible outcomes of what they do, which is an issue that itself requires discussion—the fact is that patients do not interact with healthcare professionals in a vacuum They interact with them in a particular context. That context, in turn, is limited to the resources that are available to the professionals in the particular setting in which they are engaged. Therefore the best that a hands-on healthcare professional can do is always limited by the nature of the situation in which they find themselves.

It is at this juncture that the Principle of Impossibility enters. As was pointed out in Chap. 1, one cannot have a duty to do what it is impossible to do under the circumstances that obtain.[17] Among other things, the circumstances that obtain in the delivery of healthcare include legal considerations, material resources, and the availability of appropriately trained staff. Leaving aside legal considerations—to

[17] *Supra*, pp. 14 f.

deal with these would require a separate discussion of whether a healthcare professional is ethically obligated to be aware of and to move within the legal framework that governs the jurisdiction in which they practice[18]—this means that if the material resources are insufficient in a particular regard, then the healthcare professional cannot have a duty to engage in the relevant action. In other words, it entails that if the particular type, quality or quantity of material resources that are necessary for the professional to engage in that particular action are unavailable, then the professional's obligation in this regard lapses. Likewise, if the patient's condition requires an intervention that would otherwise be in the patient's best interest but that cannot be carried out by the professional themselves because it necessitates the assistance of appropriately trained staff and that staff is unavailable, then the obligation also lapses.

With due alteration of detail, similar considerations apply with respect to the interventions that would be carried out by any AI that a healthcare professional might use. In other word, the impossibility would devolve from the healthcare professional to the tools that they use: in this case, to their AIs. Therefore the mere fact that an AI may not be able to engage in a particular therapeutic activity would not necessarily indicate that the healthcare professional who used the AI was acting unethically. However, it does entail that a properly designed AI should have subroutines that would alert their user of any such possibility.[19]

Nonmaleficence

Turning to the Principle of Nonmaleficence, it has two aspects in the context of hands-on healthcare: one with respect to the patient themselves and another with respect to other persons. The first of these is that an intervention (or lack of intervention) by a hands-on healthcare professional should not result in harm to their patient or violate the patient's competently held values. Ultimately, this is grounded in the fiduciary nature of the professional-patient relationship. The second is that an intervention (or lack of intervention) by a hands-on healthcare professional should also not result in professionally identifiable harm to other members of the society in which the patient is embedded.

---

[18] American Medical Association, "Medical laws and ethics: What to do when conflicts occur", accessed 29/09/2023 at https://www.ama-assn.org/councils/council-ethical-judicial-affairs/medical-laws-and-ethics-what-do-when-conflicts-occur

[19] Two further questions, however, arise in this connection: first, whether the healthcare professional who was faced with such a situation would have an ethical obligation to inform their patient that such was the case; and second whether, if the professional was working in an institutional facility, the professional would have an ethical obligation to inform the administration of this fact so that the latter could take appropriate steps to ameliorate this situation. The answer to these questions, in turn, is functionally related to the issue of whether healthcare itself is construed as a right or as a commodity. However, since these are questions that centrally concern the ethics of healthcare professionals and only incidentally touch the subject of AIs, further discussion of this will be put off to another occasion.

This, in turn, means that a properly designed AI that is used by a hands-on healthcare professional should not only indicate to their user what the likely outcome of an intervention or of a failure to intervene would be for the particular patient, but also what the likely outcome of this would be for the population at large.

The latter requirement, however, could not be met if the AI did not have access to data that identified the general health profiles of the population at large. Moreover, it would require that these data would allow it to evaluate the population's susceptibility to being harmed by the condition or treatment of the particular patient. In fact, it may require more than this: It may also require that the AI have access to relevant genetic data because the remainder of society may be immune to or be susceptible to such harm for genetic reasons.[20] All of this, of course, entails that a properly designed AI would have to be able to access external data bases that contained the relevant information. And this, once more, would bring into play privacy and security concerns that were mentioned above.

The Logic of Hands-on AIs

Returning, now, to the question of how an AI would have to be designed to assist a healthcare professional in ethically making a diagnosis, prognosis or treatment recommendation, it may be appropriate to take a closer look at how an AI actually functions in purely logical terms.

Data Streams

An AI, being electronic in nature, can only receive input that is in electronic form. In this respect, therefore, all input into an AI is the materially same: It differs only with respect to the origin and encoding of that input, where each type of instrumentation that originates a datum may encode that datum in a way that is different from the data that are generated by other kinds of instrumentation. Alternatively, when these do encode the data in the same way, the AI itself encodes it differently upon entry so as to keep the data distinct. Accordingly, the data that an AI uses differ in their general encoding with respect to the type of data that they are. Therefore, unless the AI's overall computational framework contains a provision that makes it possible for the AI to relate these types of encoding to one another, the data may well be received by the AI but they will remain distinct in the AI's functioning.

Consequently, if a particular type of data input that an AI can receive is called a data stream, then different types of health data belong to different data streams.

[20] Metcalf CJE, Ferrari M, Graham AL, Grenfell BT. Understanding Herd Immunity. *Trends Immunol.* 2015 Dec;36(12):753–755.

Since the data from one data stream cannot be related to the data from another data stream unless the AI's overall computational framework contains provisions to that effect, this means that an AI cannot relate health care data of one type to health care data of another type unless the AI's overarching computational framework contains the necessary provisions. Trivially, this means that if an AI's designer intends that the AI be able to correlate the data from different data streams, the designer must design this ability into the AI's overarching computational framework. Without this—as was said—a datum may well be a datum for the AI because the AI is electronically capable of accepting the relevant data stream, but the AI would be unable to interrelate the various types of data to perform a complete diagnosis.

If the totality of data in a given data stream are called a data field, then every datum in a given data field stands in a logical relationship to the other data in that field because they belong to the same data stream. This may either be with respect to its particular nature—for instance, a Fitzpatrick skin type within the data stream of skin data—or with respect to how that nature is expressed—for instance with respect to the degree that the datum is expressed, as for example degree of coloration.[21] Therefore every datum in any given data field and any given data stream has a logical association with all other data in the field and in the data stream to which it belongs in virtue of its nature and strength.

At the same time, every patient datum is related to every other patient datum simply because it is part of the totality of data about the patient, their different natures and strengths notwithstanding. This means that if an AI is to be able to perform a diagnosis, render a prognosis or to engage in an appropriate type of intervention, it must be able to correlate these different types of data with each other. In other words, it means that not only must it be able to correlate the data that are within a given data stream with other data in that data stream but also with data from different data streams. However, because the different types of data are embedded in different logical streams, an AI will actually be able to do so only if it contains appropriate subroutines to that effect.

Ideally, these would be based on standard professional data correlation protocols that are established by relevant authorities in the respective healthcare fields. However, as was pointed out before, a professional may wish to depart from standard protocols. For instance, they may suspect that while the data from a given data stream are not correlated with data from other data streams by the standard correlation protocols, such a correlation may in fact be significant. However, in order to do so—for instance, in order to explore the possibility of an innovative approach to diagnosis, prognosis or treatment of a particular condition or array of symptoms—the professional would have to transcend the data-need domain that was standardly associated with their profession. This, in turn, has ethical implications: For, by transcending their standard data-need domain, they would be going beyond the informed consent that the patient initially gave to their interaction, for this consent was on the

[21] Kechichian E, Ezzedine K. Vitamin D and the Skin: An Update for Dermatologists. *Am J Clin Dermatol*. 2018 Apr;19(2):223–235.

unspoken and understood assumption that the professional would only do what was standard. Therefore if the professional were to use an AI to this effect, not only would the AI's design have to contain a provision that would allow it to do so, its use would also have to be structured to deal with these same concerns.

Deep Learning AIs: Some Ethical Considerations

As to the provision that would allow it to do so, such an AI would have to be able to engage in what is sometimes called deep learning[22]—i.e., in a process in which it associates different levels or hierarchies and types of data and correlates them with treatment outcomes as well as with published data in a manner that emulates human reasoning. That, however, not only raises the issue of data-need domains that was indicated above, it also raises the issue of whether a patient's initial consent for the use of such an AI would be ethically sufficient. And quite aside from the issue of consent, it once more raises the issue of privacy with accompanying ethical implications.

More specifically, the consent that the patient gives to the healthcare professional is on the tacit—and sometimes even on the explicit—understanding that the professional will or may use certain tools in the performance of their actions. However, if a healthcare professional uses an AI that is capable of deep learning and that is capable of independent functioning once it was employed—for instance, by performing a diagnosis, developing a prognosis or making a recommendation—then the AI essentially performs an activity that the professional would themselves perform. Arguably, this changes the ethical picture.

That is to say, hands-on healthcare professionals normally use tools like x-ray machines, magnetic resonance imaging machines or stethoscopes, where these provide the professionals with data that they then interpret. However, when the professionals use a deep learning AI, then that AI essentially performs the task that the professionals themselves would be perform. Consequently, such AIs effectively function as substitutes for the professionals, the fact that they do not have a human nature notwithstanding. Arguably, therefore, not merely would the attending professionals ethically need the patient's consent to engage in a particular type of action, they would also be required to obtain specific consent for the use of such an AI.

To repeat, this is not because such an AI is more complicated than an x-ray machine or even than a magnetic resonance imaging machine. The complexity of a machine does not change its nature as a machine. It is because such an AI essentially executes a function that otherwise a person would perform: namely, it interprets rather than merely correlates the relevant data and it makes a recommendation.

[22] Chan HP, Samala RK, Hadjiiski LM, Zhou C. Deep Learning in Medical Image Analysis. *Adv Exp Med Biol.* 2020;1213:3–21; Syeda-Mahmood T. Role of Big Data and Machine Learning in Diagnostic Decision Support in Radiology. *J Am Coll Radiol.* 2018 Mar;15(3 Pt B):569–576; Bizopoulos P, Koutsouris D. Deep Learning in Cardiology. *IEEE Rev Biomed Eng.* 2019;12:168–193.

Therefore, it effectively acts as a substitute for the professional, and thereby it essentially functions as another person. Arguably, therefore, it is ethically appropriate for the professional to indicate not only that they will use such an AI, but also to request the patient's consent for their usage.

Moreover, if a particular kind of professional was to engage in non-standard cross-stream data correlations, their access to patient data would fall outside of the data-need domain that was standardly identified as appropriate for that type of professional. The same thing would hold true of an AI if it was capable of deep learning in the above sense of that term. It also would have to be able to transcend the standard data-need domain that was characteristic of the type of professional who employed it.

This, in turn, implies one of two things: either a deep-learning AI could not be designed with a specific data-need domain, or it would have to be designed with subroutines that would allow it to disable the limits of the data-need domain that would normally be associated with the profession of its user. In either case, however, that would raise the question whether the use of such an AI would require patient consent.

The Issue of Consent

At first glance, it might seem that this would be unnecessary simply because the patient would have consented to the patient-professional interaction when they presented for healthcare and had agreed to be treated. If that is ethically valid, then that agreement—i.e., that consent—would be have been preceded by the physician indicating what they were going to do, and this would include the methods or tools that are standardly used in that connection. After all, this is entailed by the very concept of informed consent.[23] Arguably, therefore, once the patient has agreed that the physician might follow standard procedures, then no special consent for the use of any type of AIs would be required. Special patient consent would only be required if the tools that the physician used differed from those that are standardly used. Consequently, it would seem to follow that no special notification or consent from the patient regarding the use of deep learning AIs would be required either, because deep learning AIs are merely tools. Their use would be covered by the patient's initial consent.

To argue thus, however, would be to ignore the relevance of the preceding considerations relative to deep learning AIs. It would also be to ignore an important aspect of the inception of a healthcare professional-patient relationship that was mentioned before: namely, that it occurs in a particular context.

---

[23] Cocanour CS. (2017). Informed consent-It's more than a signature on a piece of paper. *Am J Surg.*; 214(6):993–997.

To put it first with reference to the professional-patient interaction: The context in which such a relationship is incepted sets the stage for what would be involved and the data that would be required in order for that relationship to progress. It therefore includes more than merely the availability of resources: it also includes a tacit understanding on part of the patient that what data the healthcare professional will access is functionally determined by the kind of healthcare issue in question. Furthermore, irrespective of whether the patient was familiar with the standards that have been promulgated by the relevant professional authorities, the patient would be entitled to assume that the professional will be adhere to such standards.

Moreover, patients generally assume—and ethically are entitled to assume—that if the professional were to depart from these standards in any way, an informed consent to do so would be sought by the healthcare professional. So, for instance, a patient would be entitled to assume that data about whether they have a partner, how many children they have and who would function as their substitute decision maker if that need should arise would be irrelevant and fall outside of a professional's data-need domain if the professional was treating them for a sunburn, since the likelihood that the patient would become incompetent and require a substitute decision maker would be minimal. It would be a different matter if the professional-patient interaction involved the likelihood that the issue of substitute decision making might in fact arise: for instance, if the professional was treating them for a cardiac problem. The patient might become incompetent and be unable to make a decision, but a decision had to be made. Consequently, the professional's data-need domain would extend to include such information, and querying the patient or accessing the patient's EPR to this effect would also be considered appropriate. Likewise, according to standard professional protocols, data about a patient's psychiatric history would not be considered relevant if the professional-patient interaction was limited to the treatment of a broken foot. Therefore, if the treating professional was to query the patient in this regard or were to seek out such a datum in their EPR, this would also be non-standard and ethically would also require distinct patient consent. To repeat, then, if the professional was to depart from the data-need domain that was standardly associated with their profession and from the protocols that were standardly employed on that particular occasion, then the patient's consent would ethically be mandated.

When these considerations are applied to the use of deep learning AIs, they find an obvious parallel. As was indicated above, a deep learning AI departs from standard data usage protocols not simply because it trolls through a patient's whole EPR with respect to all types of health data at all levels and correlates these—possibly even in non-standard ways—but also because it accesses external data bases and connects the relevant data that are contained in them with the data in the EPR. In principle, therefore, consent for the use of such an AI would ethically be required: Not simply because, arguably, it essentially functions as a substitute for the professional but also because its actions would deviate from standard data-need domains of the attending professional. Therefore, its action would not be covered by the normal type of consent that is given to the professional with respect to the particular kind of intervention in question. After all, that consent would be based on the assumption that standard data-need protocols would apply.

The Issue of Bias

Some General Points

The discussion so far has proceeded as though both the creation of an EPR as well as the establishment of a specialty-relative data-need domain was a purely objective matter, and therefore that the design and functioning of an AI—even of a deep learning AI—followed purely objective parameters. It also has proceeded on the tacit assumption that the very identification of data—more specifically, what constitutes a datum—was a purely objective matter. Consequently, it may have given the impression that even the standard data protocols that are associated with a given kind of intervention are purely objective in nature. And this, in turn, may have fostered the impression that the design of an AI—including that of an AI that was capable of deep learning—was also a purely objective matter that involves only the identification and correlation of facts.

That assumption, however, is only partially valid. Thus, while what constitutes a datum or a fact may be an objective matter absolutely speaking, what constitutes a datum or a fact in real life—i.e., from an actor's or observer's point of view—is a relative matter. Among other things, it depends on the sensory modalities through which the observer accesses the world. Thus, a human observer who does not have a sensory modality that is sensitive to x-rays and who observes the world only through their biological senses would not naturally treat the x-ray characteristics of an entity as data because these would not register on them as observer. They would only register on them if the x-ray data are translated into a modality to which they had access: for instance, if they are translated into something like visual data that they as human observer could see. Therefore, while in and by themselves they would be data or facts, they would not be data or facts for the observer except under such conditions.

Not only that: even when the data are translated into a modality to which the observer has access, whether the characteristic in question would even register on them depends not only on the acuity of their sensory system but also on the strength with which the characteristic expressed itself. In other words, the sensory system may have a threshold level that a characteristic must reach before it would be registered.

Moreover—and quite independently of this—even if the relevant quality or characteristic met the instrumental threshold level that their system could register, and even if a datum was translated from the instrumental modality into a datum of the sort that the observer could apprehend, whether they would actually apprehend it is functionally determined by the observer's current processing scheme.

That is to say, even though a sensory system that is sensitive to x-rays and a particular x-ray input crosses the required threshold and the relevant datum is translated into a modality to which an individual has access, in order to be registered and become a datum for the observer, the observer's processing system must treat the datum as being of a kind that is relevant for the observer either at that juncture or in general. In other words, the observer will treat it as a datum only if it stands in a certain type of relationship with the other data that make up the array of data with which the observer is currently dealing. If that is not the case, then the datum will not be significant for the observer. In effect, it will not be a datum for the observer at all. Colloquially speaking, it will be ignored. To give an example of this from ordinary life, while a human driver's sensory system may register a sign that stipulates a particular speed limit, this will not register on the driver if they are focusing on the conversation that they are having with a passenger. Consequently, the diver will ignore it as a datum. This does not necessarily mean that the speed limit would not in some way register on the driver, for the driver might automatically adjust their speed accordingly. However, it will not register consciously, and in that sense it would not be a conscious datum.

When this is translated with respect to hands-on healthcare professionals, it means that what they treat as a health datum not only depends on whether the professional has access to the required type of sensory modality or to its translation into a modality to which the professional has access—sight, sound and even touch are good examples—it also depends on the professional's conceptual framework. Thus, while an oral and maxillofacial surgeon may be aware of the visual acuity of a patient or of the visual acuity data that are contained in a patient's EPR, that sort of datum will not be significant for the surgeon and hence will not be treated as a datum in the array of data and of facts with which the surgeon is consciously concerned.

However, even when professional conceptual framework treats a particular sort of datum as generally being relevant, the relevance or weight that a professional actually attaches to the item may vary from professional to professional. Thus, one surgeon may treat the ethnic extraction of a patient as a relevant datum—for instance, whether the patient was of Inuit, Scandinavian or Asian extraction—whereas this may not play a role for another surgeon. Therefore, the first would not ignore such characteristics or such a datum that was contained in a patient's EPR and would arrange their treatment protocols accordingly, whereas the second professional would ignore it and would proceed in a different way. In other words, what is here involved is what may be referred to as the issue of valuational bias.

The Identification of Data

In order to address this further, it may be appropriate to consider more closely how the use of concepts leads to the identification of data, and how a bias maybe involved in the process of acknowledging data in a given situation or state of affairs.

Thus—as was indicated above, and once more speaking purely logically—the world that presents to an observer presents as a totality of possibilities of distinction. The observer becomes cognitively aware of the world by identifying these possibilities and by grouping or arranging them on the basis of the concepts that they have. Each of the possibilities that are thus identified potentially constitutes a datum for the observer, and each aspect of their arrangement potentially constitutes a fact for them. That is why individuals with different conceptual frameworks experience the world differently: more specifically, why the world presents different sets of data to them and consists of different facts.[24] An example that illustrates this is that the world of an infant is different from the world of a sophisticated physicist, and what constitutes a datum for the one will not necessarily be a datum for the other. Consequently, their world consists of different facts. Likewise, what constitutes a fact or datum for a research chemist is not necessarily a fact or datum for a hands-on medical professional, a psychiatrist, etc.

Moreover, there is a difference between concepts that identify or pick out aspects of reality that are grounded in non-relational qualities or properties and concepts that identify or pick out relational qualities. Thus, the concept of mass picks out a non-relational property, whereas the concept of being-greater-than picks out a relational property. Moreover, relational concepts may either identify relationships that are grounded in the material natures of their relata or they may identify relationships that are grounded in non-material aspects of their relata. Thus, being-harder-than is grounded in the material natures of their relata, whereas being-smarter-than is grounded in non-material aspects of their relata. Finally, some non-material relational concepts imbue what they structure with action potentials. Values fall into this category, as do the principles of ethics. Therefore different ethical perspectives and different value systems will lead to different judgements about what would be an appropriate action and about what would be considered a fact.

When these considerations are applied to the world of hands-on healthcare practice they explain why, when an intervention is considered by a professional who subscribes to an overarching deontological ethical perspective, their activities will be structured in accordance with the Principle of Autonomy and the Principle of Equality, and that these will play a fundamental role in the conceptual framework that they use in their practice. On the other hand, this is not the case if the professional subscribes to a utilitarian ethical perspective, which is based on the Principle of Utility. Consequently, how they see a patient as well as the protocols in which they respectively engage will be different and will have a different orientation. Moreover, whether in fact a datum is acknowledged as a datum and what significance is attached to it is also functionally determined by the values to which the relevant individual subscribes. Therefore, even when healthcare professionals subscribe to the same conceptual framework and to the same ethical perspective, how they see the same patient may differ and involve different action potentials.

---

[24] Ahearn L. M. (2021). *Living Language: an introduction to linguistic anthropology*. (third ed.) Wiley-Blackwell; Kennison S (2013). *Introduction to language development* (1 ed.). Los Angeles: Sage.

Analogous considerations apply with respect to the design and functioning of an AI, both in regard to the data that it handles as well as how it handles them computationally. Therefore, irrespective of how effectively an AI may function in a technical sense, the computational gradients that determine this may be predisposed in a particular direction since the ethical principles on the basis of which these are designed will determine that the AI will operate in a particular manner. The same point applies to the values that a designer uses when designing the subroutines of an AI. These values will impose gradients on how the relevant data will be handled. Thus, values such as those that are connected with ethnicity, age or gender present such a factor. In effect, they constitute a bias.

Consequently, the design and the functioning of an AI may be biased in a particular direction. Ideally, that should be prevented because it may lead to an unethical use of the AI in question. Arguably, however, there is no way to prevent this. The very fact that the designers of AIs are human persons and that they design the structure and functioning of AIs by using their conceptual and valuational frameworks makes this inevitable. Therefore, it seems that bias is inherent in the design and construction of all AIs that are used by hands-on healthcare professionals even when the professionals are not operating in a deliberately biased fashion. The ultimate question, therefore, is not whether bias as such can be avoided, but whether a particular bias is ethically and valuationally defensible.

AIs, Telehealth and Ethics

Another series of issues—and one that was hinted at above but that was not fully addressed—concerns how hands-on healthcare professionals interact with their patients when the two are not in the same location.

That is to say, in some cases that interaction occurs in a direct interpersonal manner whereas in other cases it occurs in an electronically mediated manner by means of what is sometimes called eHealth or telehealth. It involves the use of digital information and communication technologies to manage healthcare issues and to provide healthcare services remotely. This is an emerging method of providing health care especially in geographically extended areas where, for various reasons, it would otherwise be impossible, too difficult or too expensive not to use electronic means for providing such care. The technologies may also include computers and mobile devices, such as tablets and smartphones, and the interaction may be remotely provided from a medical office or even from a mobile van. Furthermore, it may involve such things as telesurgery and telediagnosis, and all of these may involve the use of AIs. So, for instance, in telesurgery the healthcare professional manipulates devices whose functioning is transformed into electronic signals that are then translated into an AI-powered robotic device where the patient is located, and it then performs the actual operation.

Telediagnosis, in turn, occurs in two varieties: In one sort of telediagnosis the patient is connected to electronic devices that transmit the relevant data to the distant healthcare professional who then performs a diagnosis, possibly using an AI to assist them. In another sort of telediagnosis the patient is connected to electronic devices that provide an AI with the relevant patient data, and it then performs a diagnosis.

The question, of course, is whether the use of AIs in any sort of telehealth has new ethical implications. In a sense, of course, this is not the case. Thus, privacy and security as well the qualitative nature of the AIs that are used are issues that also arise in the hands-on setting. However, the situation would be different if the AIs that were involved in telehealth functioned as autonomous agents: For instance, if they were independently accessible to patients like a ChatGPT[25] and functioned as Artificial General Intelligences (AGIs)[26] without the instructions of an attending healthcare professional. In such a case, the AIs would play a distinct role from the sort of role that they play when they are used as tools by hands-on healthcare professionals and even when they are used to assist in developing a diagnosis or a prognosis. They would then essentially function like independent healthcare professionals in their own right and would seem to be independent agents. If this was the case, then the ethical issues that would arise would parallel those that arise for human healthcare professionals themselves. Arguably, therefore, distinct consent by patients would ethically be required. However, since AGIs do not currently exist, further discussion transcends the scope of the present chapter and will be put off until Chap. 6. That chapter will be wholly devoted to an analysis of the ethical implications of this possibility being actualised.

Conclusion

The focus of the preceding analysis has been ethical considerations that arise in the design and use of AIs in the hands-on healthcare setting. However, it has ignored the fact that this healthcare is frequently provided in an institutional context, as well as the fact that both the hands-on and the institutional contexts are embedded in an overall societal context that is regulated by policies that determine how healthcare itself is construed and how it should be provided. In both of these larger settings AIs may be involved in the decision-making.

To some degree, the ethical issues that arise in institutional settings overlap with the issues that arise in the hands-on setting, because the healthcare that is delivered in the institutional setting is actually delivered by hands-on professionals. Likewise,

---

[25] Dave T, Athaluri SA, Singh S. (2023). ChatGPT in medicine: an overview of its applications, advantages, limitations, future prospects, and ethical considerations. *Front Artif Intell*. 4;6:1169595.

[26] Fei N, Lu Z, Gao Y, Yang G, Huo Y, Wen J, Lu H, Song R, Gao X, Xiang T, Sun H, Wen JR. (2022). Towards artificial general intelligence via a multimodal foundation model. *Nat Commun*. 2;13(1):3094.

the strategies that are promulgated at the social policy level are translated into empirical facts by hands-on healthcare professionals. However, this notwithstanding, both the institutional as well as the societal context bring up ethical issues that affect the design and use of AIs in ways that are different from those that arise in the hands-on setting. The next two chapters are devoted to a discussion and analysis of how this is the case.

Chapter 4
AIs and Institutional Administrators

Abstract The ethical issues that arise in the design and use of AIs by institutional healthcare administrators constitute the focus of the present chapter. Particular attention is paid to the difference between healthcare construed as a right as opposed to a commodity, and to the implications that this has for the nature of the AIs that institutional administrators might use in these distinct settings. It concludes with a few considerations about the limits that are inherent in the use of AIs when the institutional setting is located in a mixed right/commodity oriented setting.

Introduction

The discussion in Chap. 3 focussed on ethical issues that arise with respect to the design and use of AIs by healthcare professionals in the hands-on setting. In passing, it was suggested that while to some degree these overlap with those that arise with respect to the design and use of AIs by institutional healthcare administrators, the latter face issues that find no real parallel at the hands-on level even though both types of individuals ultimately deal with the delivery of healthcare.

Thus, institutional administrators have the responsibility to structure and manage the protocols that determine how the institutional setting that they administer works, which means that they must integrate hands-on protocols with organizational protocols. This in turn—as will be explained in a moment—means that the AIs that they might use when executing that role have to be able to deal with some of the issues that arise at the level of hands-on healthcare as well as with issues that only arise at the organizational level.

Moreover, although the resources that institutional administrators have at their disposal are limited, just as they are for hands-on healthcare professionals, and therefore that the Principle of Impossibility is involved in both types of cases, the Principle is involved in a different way. In the context of hands-on healthcare it limits what individual professionals can offer their patients on a one-on-one basis. By contrast, at the institutional level it affects an administrator's ability to develop

E.-H. W. Kluge, *The Ethics of Artificial Intelligence in Healthcare: From Hands-on Practice to Policy-making*, The International Library of Ethics, Law and Technology 25,
https://doi.org/10.1007/978-3-031-72837-2_4

protocols that determine not only what kinds of patient-oriented healthcare the institution can provide, but also what resources the institution can provide for the professionals whom it employs or to whom it grants admitting privileges.

To be sure, while this ultimately plays out at the hands-on level with respect to what the institutional healthcare professionals can provide for their patients, at the administrative level it not only affects the nature of the patient-professional interactions as these occur at the hands-on level, it also affects the support services that the institution itself can defray in order to be able to function as an establishment. Moreover, it also affects such things as the number of patients who can be admitted in order to keep the institute functioning in an appropriate and efficient manner. Ultimately, this means that the AIs that the institutional administrators might use in carrying out their task have to be able to handle professional as well as healthcare and financial data.

To expand on this: Since institutional administrators are responsible for how the institutional framework that they administer is structured and functions, they also bear responsibility for those aspects of the institutional framework that are not directly concerned with the delivery of healthcare but that nevertheless make that delivery possible. Matters such as maintenance and finance are here implicated, and even the protocols that determine how the various types of individuals who make the institutional framework function interact with each other also have to be managed. And while dealing with these issues is facilitated by the use of AIs, the AIs that might be used in this connection are fundamentally different from those that are used by the healthcare professionals at the hands-on level and they have to span a wider informatic domain.

Moreover, while the ethical framework to which the institutional administrators and the institutionally situated healthcare professionals subscribe should be the same, how this framework functions is different. Thus, the ethical considerations that govern the activities of institutional healthcare administrators have to span both healthcare related issues that centre in the fiduciary nature of the hands-on healthcare professional-patient relationship as well as issues that fall more or less into the realm of business ethics. This does not mean that the business aspect is simply one of profit-and-loss for stockholders—although that may become relevant if the overall framework in which an institution exists considers healthcare to be a commodity—but in the sense of corporate responsibility. This also has implications for the kinds of AIs that the institutional administrators might use in carrying out their task, since the kinds of data that are involved in a corporate context are different from those that are involved in the context of hands-on healthcare. And while privacy and security issues arise in both types of settings, what these issues are and how they arise is different. Consequently, once more, the AIs that are used to deal with them have to be structured in different ways.

Furthermore, while it may be useful for the administrators of healthcare institutions to be antecedently qualified as a hands-on healthcare professional, this is not strictly necessary. While that would provide them with greater insight into the issues that face the hands-on professionals that work in the institutions that they administer, this insight can also be acquired by means of appropriate training and adequate communication. To illustrate this by means of an analogy, while experience as a criminal may be useful for lawyers in order to be able to deal with the issues that are faced by the criminals whom they represent, or for judges when adjudicating how a criminal's transgression should be handled, having had that sort of experience is not necessary either for lawyers in order to be able to represent their clients in an appropriate manner or for judges to make even-handed decisions. With due alteration of detail, then, the same point applies to healthcare administrators in the institutional settings.

As well, the AIs that the institutional administrators might use in carrying out their mandate would not necessarily require access to the same data or deal with the data in the same way as those that are used by the hands-on healthcare professionals. And even when they do access the same data—and that may well be appropriate when it comes to such things as quality assurance, for which the administrators also carry responsibility—they would not necessarily be structured with the same algorithms as those that structure the AIs of hands-on healthcare professionals.

Moreover, none of this means that institutional healthcare administrators could ignore the ethical considerations that arise for hands-on healthcare professionals that were outlined in the preceding chapter. After all, as administrators they have to establish and manage the framework within which the healthcare professionals who are employed by the institution or who have admission privileges carry out their work. Therefore—as was said—the administrative framework that they devise must integrate the ethical concerns that arise for these professionals with the administrative and business concerns that arise with the running of what is essentially a corporate body. Consequently, the tools that they have to use in discharging this task must also be able to deal with these issues, and the AIs that they use—if they use AIs at all, rather than dealing with this task on their own and in an unaided manner—must have, so to speak, a foot in both camps.

Allocation

To illustrate this further, one of the major issues that institutional healthcare administrators face—and, incidentally, one where they would find appropriately designed AIs particularly useful—is the development of an allocation framework that governs how the limited resources that are at their institution's disposal should ethically be deployed.

This task has several aspects. It involves a decision about what kinds of healthcare services the institution will provide—i.e., whether the institution will deliver all of the types of healthcare services that are currently recognized as such, including those that fall under the rubric of alternative healthcare—as well as a decision about the extent and method of that delivery. As well, it will involve a decision about whether the delivery of healthcare will be localized and involve only direct interpersonal contact between the healthcare professionals and their patients, or whether it will also include mediated interventions such as those that are used in eHealth and telemedicine.

Moreover, it also includes the task of determining the allocation framework that will be followed when several patients simultaneously present for the same treatment that the institution offers but when it is impossible to deal with them at the same time. And the allocation framework that has to be developed in this connection will have a different nature when healthcare is assumed to be a right as opposed to a commodity, even though both may be designed in accordance with the same ethical principles. Moreover, the allocation framework that the administrators put in place has to be applied consistently at all levels in that institution. This, in turn, makes the use of an appropriately structured AI that deals with allocation particularly useful.

Healthcare Construed as a Right

General Considerations

At this juncture, it may be appropriate to indicate some of the implications of this when healthcare is construed as a right as opposed to a commodity, since that would ultimately determine how an institutional administrator would have to proceed when developing the allocation protocols that hold sway in their institution and how the AIs that they use would be designed and function.

Thus, when healthcare is construed as a right, then the funds to which a healthcare institution has access in order to defray its expenses will be provided by the society that underwrites that right. This, of course—as will be discussed more fully in the next chapter—means the these funds will inevitably be limited: Not only because the society's resources are themselves limited—no society disposes of an infinite amount of resources—but also because the society's resources also have to be distributed among the other kinds of services that the society is obliged to provide for its members. Among other things, legal and educational systems also have to be funded. So is an effective infrastructure, because without such an infrastructure the society cannot develop and support the activities that will provide it with financial resources in the first place. Therefore, as was said, the amount of resources that a society that construes healthcare as a right can make available to the administrators of a healthcare institution will inevitably be limited. Consequently, resource allocation will play a crucial role in an administrator's decision about what healthcare services the institution that they administer will provide and how it will provide them.

However—and quite aside from these overarching considerations—an institutional administrator must translate the limited resources that the society puts at their disposal into allocation protocols that will function not only in an ethically appropriate but also in a practically feasible manner. These two conditions are not unconnected, for what is ethically appropriate is not necessarily what is practically feasible, and *vice versa*.

That is to say—and continuing on the assumption that healthcare is construed as a right—while it may in principle be ethically appropriate for a healthcare institution to offer transplantation services or interventions such as neurosurgery or pediatric intensive care, principles are one thing and reality is another, and the Principle of Impossibility here comes into play. The fact is that these types of interventions are resource-intensive not only in a material sense but also in terms of manpower, and the AIs and informatic services that would be necessary to support such undertakings would also have to be kept available and operational as well as be updated on a continual basis. Consequently, providing high-profile healthcare interventions in a qualitatively appropriate manner would reduce the amount of resources that would available for providing the types of more common healthcare services that the institution might also provide. All of this has important implications for the nature and design of any AIs that the administrators might use when making the relevant decisions. Minimally, the AIs would have to be able to correlate factors that involve financial, manpower and quality-of-care considerations.

It may therefore seem that the considerations that would ultimately influence an institutional administrator and that would ground the operational routines of any AI that they might use would essentially be financial in nature.

However, to view the allocation task that faces institutional administrators in purely financial terms would be too short-sighted. It would be to ignore another factor that was already mentioned, but that deserves explicit consideration. More specifically, it would be to ignore the ethical implications of the various domains of information that would have to be accessed when making the relevant decisions. As well, it would be to ignore the operational aspects with which an AI would have to deal when facilitating such decision-making.

More specifically—and putting it only in terms of the AIs that might be used in this regard—not only would the AIs have to be able to access the costs that would be associated with the relevant interventions insofar as the institution was concerned, they would also have to be able to access relevant data on the effectiveness and efficiency profiles of the interventions themselves. After all, it is only in this way that it could be determined whether the institution should offer the kind of healthcare intervention in the first place. For, ethically, a healthcare institution should not offer an intervention if it could not provide the resources that would allow its professionals to meet the current standards. That would be to interfere with the professionals' fiduciary obligation.

However, in order to be able to do this, the AIs would have to be able to access not only the EPRs of the affected patients and the external data bases that identified what interventions would standardly be expected, but also the data bases that identified the kind of outcome that could reasonably be expected if the institution provided the relevant kind of intervention. As well, it would require that the AI have access to data that dealt with the efficiency profiles of the healthcare professionals who would be involved in delivering the relevant services, as well as the institutional service-profile that would be required in the delivery of such intervention. It would then have to be able correlate these factors. Only in that way would it be able to provide the institutional administrator with a useful basis for making the relevant decision.

The issue extends still further. The discussion so far has focused on matters that arise in the institutional delivery of health care with a focus on healthcare itself. However, the issues that an institutional healthcare administrator has to consider include more than only the factors that are more or less directly involved in the actual delivery of healthcare. They also include matters that centre in the legal reaction that the delivery of that care may trigger in the recipients of that care: namely, in the patients. In other words, it also extends into the field of law because this might have budgetary—and hence operational—implications.

That is to say, an administrator would have to be aware of the possibility that legal actions might be taken by patients against the institution itself and against the professionals who work in that institution. This is because the patients might be dissatisfied with the overall institutional framework in which they received their care, with the conduct of the professionals who provided that care or with the outcome of the interventions that they received. Therefore they might initiate legal proceedings in this regard. That, in turn, means that if an institutional administrator used an AI to deal with such matters, the AI would also have to have access to relevant legal cases.

Of course, an administrator might want to deal with such issues on their own and without using an AI. Alternatively, they might want to employ an appropriately qualified legal advisor. However, if they chose the latter, then not only would that be a financial item in its own right and therefore would have to be subtracted from the resources that would be available for the delivery of healthcare, it would also mean that the administrators would have to devise some way in which the legal advisor could have access to the relevant records not only of the institution and of the healthcare professionals that were involved, but also of the patients who brought such actions. For, the mere fact that a patient advanced a legal claim would be insufficient to establish the validity of the patient's contention. Therefore, in a legal action the institutionally stored patient records that contained the details of how the patient was actually treated and what the actual outcome of their treatment was would have to be compared with the records that were adduced by the patient. This, in turn, means that privacy and security issues once more would have to be dealt with. And with this, the design of any AIs that might be used would become ever more complicated. In other words, any AIs that the administrators might use when making the relevant decisions would have to be able to straddle distinct informatic domains without violating financial, privacy, security and ethical concerns.

Hands-on Institutional Allocation Protocols

What so far has been considered under the rubric of resource allocation have been items of an essentially hybrid nature that bridge financial, professional, legal and healthcare matters. However, there is an element that also falls under the rubric of allocation but that does not cross over into matters that are not healthcare related. It is an element that is purely ethical in nature. It was mentioned above, when reference was made to the situation where several patients simultaneously present for the same treatment modality but not all of their claims can be satisfied at the same time. It is an issue that can be captured by the question "Who shall have when not all can have?"

The reason that this is something that an institutional administrator has to deal with is the fact that they are responsible not only for the efficient functioning of their institution, but also for its ethical operation. This means that the operational framework of the institution that they administrate not only has to incorporate allocation protocols that appropriately turn away certain types of patients—or instance, prospective patients who seek interventions that the institution does not provide—but also protocols that distinguish the order in which patients who meet the institution's entry requirements will be treated when not all of the patients can be treated with the same modality at the same time. The need for such a protocol is grounded not only in the fact that the number of qualified professionals that an institution can employ is limited, but also that its material resources—inclusive of space and relevant devices—are limited. Therefore, sometimes such decisions have to be made.

Moreover, the efficient functioning and ethical consistency of institutional operations requires that such decisions should not be left up to the perspective of the institute's admission personnel or of the healthcare professionals themselves, for this might vary from individual to individual. The only way to ensure that such decisions are made consistently and ethically appropriately is to put in place an appropriate decision-making protocol.

Therefore, an important task that faces an institutional administrator is to put in place an efficient and ethically acceptable allocation protocol in matters of healthcare at the hands-on level, where this protocol will determine the order in which eligible patients who compete for the same resources will be treated when they cannot all be treated at the same time. While the allocation decisions will actually be made at the hands-on level, an administrator cannot assume that the relevant staff and professionals are trained in ethical decision-making, nor can they assume that the personnel will actually make them in the same way.

Consequently, to ensure that such decision-making is both efficient and ethically valid, the administrator may well have its format incorporated into the subroutines of an AI and require that all personnel, when faced with the need to make an allocation decision, use that AI. In that way, it would be highly likely that all issues of allocation would be handled in a consistent and ethically appropriate manner.

AIs, Calculations and Ethics

However, it might seem that to argue thus would be to contradict the claim that was advanced in the preceding chapters: namely, that expert system or AIs are not ethical decision-makers. It was then argued that only persons can function in that capacity. Yet, arguably, if it was institutionally required that an AI be used when an allocation issue presented itself, then the AI would in fact be ethical decision-makers after all.

However, to argue thus is to confuse the nature of a tool with how the tool is used. The two are logically distinct. For instance, a calculating device deals with the numbers that are entered into it by handling them according to the algorithms that have been built into it. It therefore gives different answers if the numbers and the problems that are entered into it are different. These answers, in turn, may have ethical implications. However, they are not ethical in and by themselves. It all depends on how they are used. For instance, when a financial advisor uses a calculator to determine how best to invest a client's money for the client's benefit, it is not the calculations themselves that are ethical. It is their use. Thus, their use would be ethically appropriate and in accordance with the financial advisor's fiduciary obligation towards their client if they actually used these calculations to make decisions that benefitted their client. On the other hand, if the advisor were to use a calculator to determine how best to acquire the client's money and how best to use it for their own benefit, then their use that would be ethically inappropriate.

With due alteration of detail, similar remarks apply in the case of the allocation AIs that are to be used by institution's admission personnel and by its healthcare professionals when it comes to matters of allocation. The AIs would be mere tools. And while the protocols according to which the AIs functioned may have ethical implications, the actual ethics lies in the use of their calculations and not in the calculations themselves. Therefore, while the subroutines that perform an AI's calculations with respect to hands-on resource allocation may be structured in a way that attaches a particular significance to certain data, this does not make the calculations that are carried out by these subroutines ethical or unethical, nor does it turn the AI that performs such calculations into an ethical decision-maker. Just like the calculator that was mentioned above, the AI would still only be a tool or device. It is the application of the results of its calculations by their users that would or would not be ethical.[1] Therefore ethical decision-making would still in the hands of persons.

Allocation Protocols

Examining, then, the parameters that the administrator of a healthcare institution that was situated in a right-oriented society would be obliged to consider relevant and that therefore should be reflected in the subroutines of the AIs that would be

[1] Cf. Chap. 2, p. 24 *et pass.*

institutionally used in this connection, the following would play a pivotal role: consent, effectiveness, time-sensitivity, life-saving vs. quality-enhancing, hetero- vs. auto-induced, and claimed vs. not claimed. The ethical principles that would justify their involvement would be the Principles of Autonomy, Impossibility and Equality. The design of the subroutines of an AI that was mandate to be used would then have to incorporate these elements into an effective and consistent operational schema.

Consent[2]

Thus, a properly designed allocation protocol would begin with the matter of consent. That is to say, when several individuals arrive for a treatment that the institution offers but only one of them can be treated at that point in time, ethically speaking the first issue that would have to be considered is whether the individuals either directly consented to being treated, may reasonably be assumed to have consented, or that a duly empowered substitute decision-maker acting in accordance with the patient's values had provided such consent in place of the patient.[3] In other words, the Principle of Autonomy would here be determinative.[4]

The reason it would be ethically appropriate to incorporate this into an allocation protocol as an initial consideration is that not all persons who arrive at a healthcare institution do so voluntarily. They may arrive as a result of having been brought by emergency response personnel or in some other way that did not involve their own decision. However, the mere fact that they are persons and have a right to healthcare does not itself directly entail that they should be treated because—all other things being equal[5]—a right is not a duty. In other words, as persons they have the option not to avail themselves of this right; or, alternatively, they may wish to avail themselves of that right but the kind of healthcare that the specific institution provides differs fundamentally from the values to which they subscribe, and therefore they may decline the kind of healthcare that the institution offers.[6] For instance, that would be the case if a devout Jehovah's Witness undergoing severe blood loss was brought to a hospital. In such a case, the otherwise medically appropriate intervention would be to provide a transfusion. However, the Principle of Autonomy would here be engaged, and an ethical allocation protocol would determine at the outset whether including the Jehovah's Witness patient in the group that competes for the institution's blood resources would be ethically appropriate. As was said, while an

---

[2] In what follows, the term "consent" will be understood as being short for "consent or refusal".

[3] Cf. Chap. 3, p4. 5 *et pass.*

[4] Cf. Chap. 3, *passim.*

[5] Cf. Chap. 1, pp., 15 *et pass.*

[6] WMA *Code of Medical Ethics*, accessed 02/09/2022 at https://www.wma.net/policies-post/wma-international-code-of-medical-ethics/#:~:text=respect%20a%20competent%20patient's%20right%20to%20accept%20or%20refuse%20treatment.&text=not%20allow%20his%2Fher%20judgment,personal%20profit%20or%20unfair%20discrimination.&text=be%20dedicated%20to%20providing%20competent,and%20respect%20for%20human%20dignity

individual may consent to be treated, they may decline being treated in a particular way.

Another kind of situation would be one where a patient arrives at the institution, was unconscious, a duly empowered substitute decision-maker was not reasonably available but the patient had a readily available advance directive[7] that indicated that they would accept whatever palliative care was available, but that they would not want any life-saving interventions. Unless the healthcare institution could not provide palliative care—in which case the allocation protocol should indicate that the patient should be transferred to an institution where such care would be delivered—the admitting personnel should remove such a patient from the acute care stream and apply the next criterion to the remaining patients.

Effectiveness

The next criterion that an ethically designed allocation protocol would then apply is whether the intervention for which the patients competed would be likely to have a positive outcome for the patients in question: In other words, whether it would be likely to be effective. If not, then those patients for whom this was not the case should be diverted to a treatment routine that would provide them with otherwise appropriate care, and consideration should be focused on the remaining patient or patients whose treatment was likely to be effective.

The ethical reason for this, in turn, would lie in the Principle of Impossibility: If a particular right-claim cannot be satisfied, then the obligation to act in accordance with that right claim lapses.[8] This does not, of course, mean that the right to healthcare itself lapses, because that would not hold for the other healthcare that could be provided; but it does hold for the intervention in question.

Time-Sensitivity

The next criterion that would play a fundamental role when making an allocation decision is whether the intervention in question was time-sensitive: In other words, whether the treatment in question would only be effective if it was provided at that point in time. If it was not time-sensitive, the treatment of the individuals for whom this was not the case would be put off at that point in time, and the treatment of the individuals for whom this was the case it would take priority.

The ethical reason for this lies in a combination of the criterion of effectiveness and the ethics of the healthcare professional-patient relationship. The reason the criterion of effectiveness is involved is that if a particular intervention is likely to be effective only at a specific point in time, then this means that the intervention

---

[7] Arras J. Ethical issues in emergency care. *Clin Geriatr Med.* 1993 Aug;9(3):655–64.; Hong CY, Goh LG, Lee HP. The advance directive—a review. *Singapore Med J.* 1996 Aug;37(4):411–8.

[8] *Cf.* Chap. 1, p.17 *et pass.*

logically carries what may be called a time-rider. In other words, it means that if the treatment was not provided at that point in time, then it would in fact not be an actual treatment but something more in the sense of being merely a certain type of activity.

The reason that the ethics that governs the behaviour of healthcare professionals is involved is that, as was pointed out in the preceding chapter,[9] it is generally accepted that healthcare professionals have an obligation to act in the best interest of their patients. In other words, the Principle of Nonmaleficence, as incorporated into traditional as well as contemporary Codes of Medical Ethics, enjoins the healthcare professionals to do their best not to cause harm to their patients. Consequently, if the intervention was time sensitive in a particular case and the same intervention was not time-sensitive in another, the former should take priority.

Life-Saving vs. Quality-Enhancing

Once that issue has been resolved and several patients still satisfied the preceding criteria, then the next issue that would have to be decided is whether, in the case of the remaining patients, the intervention in question would be life-saving or quality-enhancing. More specifically, the allocation protocol would specify that the treatment of those patients in whose case the intervention would be life-saving would take priority over the treatment of those patients where the proposed treatment would not be life-saving but merely improve the quality of their life.

The reason for this condition lies in the very nature of the right to healthcare itself. More specifically, one cannot have a right to healthcare if one is not alive. Therefore not to engage in the life-saving intervention would effectively be to deny the right to healthcare of the patients whose condition was life threatening, whereas that would not be the case if the patients in whose case the treatment merely had a quality-enhancing effect was relegated to another and different treatment stream. To be sure, that may mean that the quality-enhancing treatment might not be as effective as it would be if it were immediately provided, but that does not change the logic of the distinction.

Auto-Induced vs. Hetero-Induced

Returning, then, to the overall structure of an ethically appropriate allocation protocol, if the criteria still did not determine who should be treated at that point in time and who should be relegated to a different treatment stream—either to one that provided the relevant treatment at a different point in time or to one that essentially provided palliative care—then the next criterion for distinguishing between the competing claims would be the criterion of responsibility: Which is to say, whether

[9] Chap. 3, pp. 51 *et pass.*

the condition in question was auto-induced or hetero-induced. If the condition was auto-induced—in other words, if it was the result of a voluntary act or resulted from a voluntary failure on their part to act in an appropriate manner—then it should be accorded a lower priority than the treatment of the individual whose condition was hetero-induced.

The ethical reason for this lies in a combination of the Principle of Autonomy, the Principle of Impossibility and the right to healthcare itself. More specifically, and as was indicated in Chap. 1, the Principle of Autonomy is to the effect that everyone has the right to self-determination. However, as also indicated, this Principle carries an important qualifying clause: namely, s*ubject to the equal and competing rights of others.*[10] Consequently, voluntary acts or failures to act are ethically permissible only as long as they do not interfere with the equal rights of other persons. Therefore if the voluntary act (or failure to act) on the part of one person results in a condition that requires a healthcare intervention but providing such an intervention interferes with the ability to provide another person who was not responsible in that regard with necessary healthcare, then the claim of the initial actor that they had a right to act (or to abstain from acting in the relevant fashion) and that this should not affect the priority-ranking of their right to be treated, would be invalid. Accepting such a claim would be to ignore the conditioning clause of the Principle of Autonomy itself.

A Qualification

However, there are some ethically important exceptions to the preceding consideration, and a protocol that did not recognize them would be ethically deficient. Thus, there are certain kinds of voluntary actions that may result in health conditions that require treatment and that therefore are auto-induced but that differ from other types of voluntary actions in certain fundamental ways.

Thus, certain occupations carry a risk of injury or of health issues that require attention. For instance farming,[11] policing[12] and healthcare professions themselves[13] fall into this category. However, the exercise of such occupations is necessary for the functioning of a society and for the very existence of a healthcare system itself. In other words, while these occupations are exercised on a voluntary basis, the auto- vs. hetero-induced criterion does not apply in their case because, despite being exercised on a voluntary basis, the Principle of Impossibility is engaged. For, if they were not carried out, then society itself would not exist as a functioning entity and could not provide healthcare in the first place.

[10] CF. *supra* Chap. 1, p. 13 *et pass.*

[11] Centers for Disease Control and Prevention, "Agricultural Safety", accessed 24/06/2023 at https://www.cdc.gov/niosh/topics/aginjury/default.html

[12] National Institute of Justice, "Officer Safety", accessed 24/06/2023 at https://nij.ojp.gov/topics/law-enforcement/officer-safety

[13] Centers for Disease Control and Prevention, "Healthcare Workers," accessed 24/06/2023 at https://www.cdc.gov/niosh/topics/healthcare/default.html

As well, not all persons who present for healthcare and who otherwise would be subject to the auto- vs. hetero-induced allocation criterion are volitionally capable. Consequently, they cannot reasonably be held responsible for their actions. Therefore, while they may themselves bring about their health condition, the notion of autonomy does not really apply them. Therefore in their case, the Principle of Autonomy itself grounds the exception.

Drawing Lots

However, it is possible that even when these distinguishing criteria are taken into account, competition for limited resources still exists because more than one individual may still meet the preceding criteria. Nevertheless, a decision has to be made, lest no-one would receive the relevant healthcare and therefore the right to healthcare would lapse in all cases. Consequently, the ethically appropriate way to structure the allocation protocol would be to assign each individual an equal chance of being treated and then to draw lots as to who will receive the intervention.

A Critique and its Resolution

At this point might be argued that this would in fact be to advantage one individual over the other, and that this would violate the Principle of Equality. Moreover, that it would also mean that the individuals who were not treated would be harmed, which in turn would mean that the Principle of Nonmaleficence would be violated. It is arguable, therefore, that ethically neither patient should be treated, for that is the only way in which the Principle of Equality would be satisfied.

This, so it may be argued, shows that it is impossible to design an allocation protocol by using a deontological ethical perspective, and therefore that it is impossible to incorporate a deontological ethical framework into the design of an AI that is used in allocation. Consequently, it seems that a deontologically based allocation protocol is ultimately unusable in practice, and therefore that an institutional administrator should adopt a different ethical perspective when it comes to designing a protocol for allocation.

However, to argue thus would be to construe the Principles of Equality and of Nonmaleficence too narrowly. More specifically, while deontological Principles apply to all persons, to construe them as only applying to natural persons is to misconstrue their ambit. For, as was pointed out in Chap. 1, the notion of personhood extends to both natural and constructive persons.[14] To be sure, societies, corporations, institutions and collectivities are constructive persons, but they are persons nevertheless. Therefore, both the Principle of Equality and the Principle of Nonmaleficence apply to them as well. And just as a natural person is constituted of parts—for instance, human beings are constituted of hearts, livers, skin and brains,

---

[14] Chap. 1, pp. 11f *et passim*.

etc.—constructive persons are also constituted of parts: namely, of the natural persons who are their members.

Also, at this juncture it is relevant to recall that a part of a natural person may be harmed in the provision of healthcare: For instance, when a cancerous growth is surgically removed by cutting open the individual's abdominal cavity and excising the relevant tissue or by performing an appendectomy, or even by providing pharmacological agents that have negative physiological side effects, etc. Such actions, however, are nevertheless considered to be ethically acceptable because notion of harm applies to the whole person, not to their parts.

When the logic of this is applied to the case at hand, it entails that even though not treating one of the patients would result in harm to that patient and therefore to a part of the constructive person that is the society, the amount of harm that would be thereby be produced for the constructive person would be less than if neither of its parts—i.e., neither of the patients—was treated. To revert back to the previous examples, this is analogous to the fact that the material harm that is produced in some medical intervention such as opening up the abdominal cavity and excising the cancerous tissue constitutes a benefit to the person of the patient. Drawing lots treats each of the constructive person's part—that is to say, treats each natural person who is a part of society—in an analogous manner. Therefore the Principle of Equality would be satisfied and the apparent conflict between the Principle of Equality and Nonmaleficence would in fact be resolved. Admittedly, it would be tragic that an individual who was not treated would suffer. That, however, would not make the allocation protocol that was outlined above unethical. After all, not everything that is tragic is unethical.

Implications for AI Design

If the preceding considerations were incorporated into the design of an AI that was imposed by an institutional administrator on admission staff and on healthcare professionals, it would ensure that the institutional allocation procedures that were followed would not only be efficient and consistent but also ethically appropriate.

To put this is more specific terms, the first issue that would be programmed into the AI would to prompt the allocation decision-maker to determine whether the patients in question were capable of giving or refusing consent to the specific kind of intervention that was at issue. If the patients were incompetent, a subroutine would advise the decision-maker to determine whether the patients had a reasonably available advance directive that indicated whether they would agree to the kind of intervention in question. If the patients were incompetent and an advance directive was not reasonably available, it would prompt the decision-maker to determine whether a duly empowered substitute decision-maker was reasonably available to give or refuse consent. If none of this was the case, it would prompt the decision-maker to determine whether it could reasonably be assumed that someone in the patients' condition would consent to the treatment in question.

If consent either on part of a competent patient or, in the appropriate case, on part of their duly empowered substitute decision-maker was forthcoming—or, in the absence of this, if a reasonably available advance directive indicated consent or it was reasonable to assume that the patient would consent to treatment—then the AI would prompt the professional to proceed to the criterion of time sensitivity. On the other hand, if consent was duly refused—or could reasonably be assumed to be refused—the AI would prompt the professional to remove the relevant patient from the competitive treatment stream and provide them with whatever other treatment was institutionally available.

The AI would then prompt the decision-maker to determine whether the treatment of the individuals was in fact time-sensitive—i.e., whether it would be effective only if it was provided at that point in time—or whether it would still be effective if it was provided at a later date. If it was not time-sensitive, the AI would prompt the decision-maker to remove the patients in whose case treatment was not time-sensitive from the treatment stream and relegate them to another stream that would provide them with whatever other appropriate treatment was available or that would provide the treatment in question at a later date.

If in each case the treatment was time-sensitive, the AI would then prompt the decision-maker to determine whether the relevant treatment was life-saving or quality-enhancing. In those cases where it was not life-saving but merely quality-of-life enhancing, it would prompt the decision-maker to remove the relevant individual from the competitive treatment stream and relegate them to another treatment steam that would provide them with whatever other treatment was available or provide them with the relevant treatment at a later date.

If in each case the treatment was life-saving, the AI would prompt the decision-maker to proceed to the next criterion, and determine for each of the remaining cases how the relevant condition came about: i.e., whether it was auto- or hetero-induced. If the professional determined that in each case it was hetero-induced, the AI would prompt the decision-maker to proceed to the criterion of drawing lots. On the other hand, if the professional determined that the condition was auto-induced, it would prompt the decision-maker to remove the relevant individual from the competitive treatment stream and relegate them to another stream that would provide them with whatever other treatment was available. If that left only one patient in the competitive treatment stream, the AI would indicate that it was ethically appropriate to treat the individual in question. On the other hand, if that still left more than one patient, the AI would prompt the decision-maker to cast lots as to who should be treated first. The so-called drawing of lots could be programmed into a subroutine of the AI.

However—and this acknowledges the exceptive conditions that were mentioned above—before doing so, an appropriate AI subroutine would prompt the decision-maker to determine whether the auto-induction was the result of an action that was integrally involved in maintaining the functioning of the healthcare services themselves, was the result of carrying out an activity that was integrally involved in the functioning of society, or the individual's condition was the result of a lack of autonomy. If it was determined that any of these was the case, the AI would prompt the decision-maker to ignore the auto- vs. hetero-induced criterion and cast lots as to who should be treated first.

By introducing such an AI into the decision-structure of their institution and requiring that all professionals use such an AI when faced with an allocation issue, the institutional healthcare administrator in a right-oriented setting that subscribed to a deontological ethical framework would ensure that the limited institutional resources would always be distributed in consistent and ethically acceptable fashion, and that the perspectives or values of the professionals would not play a discordant role in the institution's functioning.

Healthcare Construed as a Commodity

The analysis so far has been based on the assumption that the institution was situated in a society that construed healthcare as a right. However, not all societies subscribe to such a perspective. Instead, they consider healthcare to be a commodity. If that was the case, an institutional administrator would have to proceed somewhat differently, and the design of an AI that was to be used for resource allocation at the hands-on level in that institution would also have to be somewhat different.

Some Qualifying Considerations

Of course, a society's legal framework might well require that some of the considerations that were outlined on the basis of healthcare construed as a right would be binding even though healthcare in general was construed as a commodity. Some of the countries in Europe as well as in the Americas would here be illustrative examples. If that was the case, then an institutional administrator would have to structure the allocation protocol that should be followed in the institution that they administer in the same way as was previously indicated, and the AI that they might use in designing and implementing such a protocol would be structured along the same lines.

That, however, would be a matter of law, not of ethics. For with respect to ethics, some importantly different issues would arise for institutional healthcare administrators when it comes to developing a resource allocation protocol when healthcare is construed as a commodity. And even if the administrator used a deontological ethical perspective when doing so, they would still have to take into account somewhat different issues. Consequently, as was said, any AIs that they might use would also have to be designed somewhat differently, and they would have to deal with issues that differed from those that have been outlined so far.

As well, it might even be the case that the codes of ethics of the healthcare professionals who would actually would have to deal in a hands-on fashion with resource allocation in a commodity oriented setting might appeal to the same ethically based considerations that were just sketched, and that the professionals saw themselves bound by these considerations because they were involved in their codes of ethics. Moreover, that would include hands-on professionals irrespective of whether they were situated in an individual practice or in a larger institutional setting. Consequently, in order to meet what they would consider to be their ethical obligations, any AIs that they might use in this regard would also be designed along the same lines as the ones that their colleagues used in a right-oriented healthcare setting.

If that was the case, however, it would pose an almost intractable problem for the institutional administrators in a commodity oriented setting. For, it would mean that the overarching institutional framework for which they were responsible would have to be structured along commodity-oriented lines while some of the ethical issues that they would have to deal with would be centred in a right-oriented perspective. Consequently the AIs that they could use when dealing with these issues would have to be structured along essentially incompatible lines.

That aside, however, and the difference between a right- and a commodity-oriented perspective notwithstanding, there would be an overlap with respect to some ethical concerns. For instance, the privacy and security concerns that were indicated in the above analysis would still arise even though healthcare was construed as a commodity, because these are not issues that derive from how healthcare is construed but from the sphere of privacy that surrounds each person as person. Therefore the AI tools that administrators could use when designing and implementing the relevant institutional provisions would also have to be designed to deal with these issues.

AIs and Admission Protocols

The important point here is that when healthcare is construed as a commodity then, by offering healthcare services, a healthcare institution would essentially be a seller of commodities. However, it is generally understood that the purveyor of a commodity has an ethical as well as a legal obligation to ensure that the commodity that they offer for sale was "fit for the purpose". Moreover, that it is accurately described or advertised. This not only applies to the content of the offer but also to the conditions under which the commodity is offered. Therefore, the administrator of a healthcare institution in a commodity oriented setting would also be obligated to ensure that the information that was provided about the services that the institution offers was complete and accurate in all relevant respects.

At the same time, however, it is also generally understood that it is the purchaser of a commodity who has an obligation on their own part to take due care that they understand the nature of what they are purchasing and the conditions that surround their purchase. That obligation does not fall to the seller. Thus, while the seller of a car has an obligation to be truthful about what they are selling and to inform a prospective purchaser of all aspects that a reasonable person would want to know before deciding on the purchase, it is the prospective purchaser's obligation to make sure that they actually understand the information that they are given. Strictly speaking, therefore, when healthcare is construed as a commodity, the same considerations apply with respect to the healthcare that is offered by a healthcare institution.

Therefore, as was said and strictly speaking, in a commodity-oriented setting an institutional administrator would not have the obligation to ensure that prospective patients actually understand the information that they were given. Consequently, any AI that they might put in place to ensure that admission occurs in an ethically appropriate and consistent manner would not have to contain a subroutine whose purpose was to establish that prospective patients actually understood the health services that they were purchasing. While this would be ethically required if healthcare was construed as a right and should be reflected in the AI that might be used when dealing with the issue of patient consent, this would not be the case in a commodity oriented healthcare context—unless, of course, that was stipulated by relevant overarching legal provisions.

That is to say, even when healthcare is construed as a commodity, legal provisions or professional codes of ethics may actually mandate that the purchasers of the relevant healthcare services did actually understand what it was that they were buying. If such was the case, then the AIs that the administrators would put in place for the purpose of admission and allocation would have to contain a subroutine that would require that the admission professionals determine whether the prospective patients actually understood what they were purchasing. However, as was said, that would be a matter of law and not of ethics. As to codes of ethics, it would be a matter of ethics only in the extended sense because such codes are ultimately promulgated on the basis of a political process, rather than on the basis of purely ethical considerations.[15]

Also, as a matter of law or of codes of ethics, even in a commodity-oriented setting there might be a fiduciary relationship between the individual patients who are admitted to a healthcare institution and the healthcare professionals who actually provided that care. However—and quite aside from the fact that this type of relationship would have to be justified on an independent basis—that would not be the institution's responsibility unless, one more, this was stipulated by the legal framework within which the institution operated. Consequently, it need not necessarily be reflected in the institutional protocols and in the design of any AI that the institutional administrator might use or put in place to structure the institution's operations. It would be the responsibility of each individual healthcare professional to structure their interaction with their patients in the relevant manner.

Nor—to return to the sale of commodities—does the purveyor of an item or of a service have an obligation to ensure that what they offer for sale is the best possible or the most up-to-date. What they are selling is essentially an as-is item, and it is up to the purchaser to decide whether they wish to buy it. With due alteration of detail,

[15] Riddick FA, Jr. (2003). The Code of Medical Ethics of the American Medical Association. *Ochsner J* 5(2): 6–10.

this also applies in the case when healthcare is construed as a commodity and if the institution exist in a commodity-oriented framework. Therefore the institutional administrator would not be obliged to institute relevant protocols in this regard or to have this reflected in any of the AIs that were used in evaluating the appropriateness or effectiveness of the particular kinds of healthcare that the institution provides. It would essentially be a situation of "buyer beware."

Of course, as was said, both the codes of professional ethics of the healthcare professionals that are employed by the institution as well as the legal provisions that bind the healthcare institutions as a corporate entity might well compel both professionals and institutions to follow strictures of the sort that were outlined above with respect to professionals and institutions in a right-oriented setting. Consequently, an institutional administrator might have no choice but to adjust the operational protocols accordingly and to have these parameters included in the design of any relevant AIs.

That, however, would entail that such a healthcare institution would, so speak, have to serve two masters: In other words, that the institutional administrator would be faced with coordinating potentially conflicting tasks. Furthermore, it would mean that any AI that they might use could not have a single unified operating system. The AI would then have to function on the basis of distinctly designed subroutines that respectively applied to the distinct types of situations. And while these subroutines might be located in one and the same computer, that computer could not have a single and unified operating system except in the sense that it would switch from one type of subroutine to another as the occasion demanded. However, whether and how it switched its operations would ultimately be up to the institutional administrator.

Variety of Services

The difference between the two kinds of approaches to healthcare extends further, and includes such issues as the variety of services that would be offered both at the policy and at the hands-on level. Once more, this derives from the fact that the primary obligation of the administrators in a commodity-oriented setting would not be towards the patients with whom the institution deals but towards the institution's stockholders, whether that was society itself or the private owners of the institution.[16] Consequently, unless the stockholders determined that the institution should function at a loss—for instance, if the stockholders were society itself and the society accepted that the healthcare institution might operate at a loss, or if the institution was funded by a non-profit benevolent society that would be willing to fund it on a similar basis—the institutional administrator would be ethically obliged to

---

[16] USENIX Association, LISA, and LOPSA. (2012) *System Administrators' Code of Ethics*, accessed at https://www.usenix.org/system-administrators-code-ethics

craft policies and protocols that would be likely to ensure the viability and, if possible, the profitability of the institution in financial terms.

This, in turn, would mean that the AIs that might assist them in making the relevant decisions would have to take into account cost-benefit considerations. Therefore, such AIs would include not only subroutines that would evaluate the cost-income ratio of the relevant services but also subroutines that would evaluate the possible impact of any legal actions that patients might take on the basis of what they perceived as lack of personnel or the quality of the interventions that they received. While the latter might also figure in an AI's subroutines in a right-oriented setting, the focus there would be the welfare of the patients, whereas the focus here would be the financial implications that such actions would have on the ability of the institution to function.

Not only that: If the society in which the healthcare institution was situated did not regulate what kinds of services a healthcare institution should provide as a matter of law, and if healthcare itself was considered a commodity, then what services such an institution would offer as well as the degree to which it would offer them would be functionally determined by considerations of profitability.

At first glance, this would seem to parallel the considerations that were raised when indicating that a healthcare institution in a right-oriented setting would not necessarily offer all of what might otherwise be considered appropriate healthcare services. It would therefore seem that an institutional administrator in a commodity oriented setting would also not necessarily incorporate certain services into the delivery structure that they designed and managed.

However, although the two are related, they are not quite the same and their respective justifications are importantly different. Thus, in a right-oriented setting the Principles of Impossibility, Beneficence and Nonmaleficence would play a central role in justifying such a decision. For, in such a setting a healthcare institution would be a means that was used by society to fulfil its obligation to provide healthcare to its members. Consequently, for reasons that were pointed out before, the society would have to apportion its healthcare resources in such a way that not all healthcare institutions would provide certain services, and it would mean that not all healthcare institutions could provide the respective services in the same way. Consequently, it would be ethically appropriate for healthcare institutions in such a setting to divert the patients who needed certain services to institutions that would be able to offer them.

By contrast, when healthcare is construed as a commodity the fact that some healthcare services are resource-intensive plays a different role. In such a setting, an administrator would ethically be obliged to design the institution's operational framework in such a way that, minimally, it would recover the costs that would be involved in providing the relevant services. Consequently—and all other things being equal—they would be obliged to exclude certain services if the institution's overall profitability would be negatively affected. Consequently, the relevant calculations would have to be designed into the programmes of their AIs.

At the same time, however, this would not necessarily be the case. Thus, the relevant administrator could take a close look at the expenditure profiles of such things as cosmetic plastic surgery. This type of surgery tends to be used by well-to-do clients and tends to be very profitable for both healthcare institutions and healthcare professionals.[17] It even tends to be the case that even though the clients or patients might not be satisfied with the outcomes of such interventions, the overall institutional profit balance would still be positive.[18] Consequently, they might still decide that the institution should provide such services. As well, the institutional administrator might also decide that the institution should also offer non-standard treatments or alternative healthcare interventions, since these also tend to constitute profitable items and to find a ready market.[19] Moreover, the administrator of a healthcare institution in a commodity oriented setting might also adopt a loss-leader perspective. That is to say, they might adopt a marketing strategy where the reduced income that resulted from proving unprofitable healthcare services was set off by attracting patients to whom the institution could sell the more profitable healthcare services.

Consequently, the AIs that the administrator would use in planning such measures would have to be designed to access the relevant market reports and to incorporate those data into their calculative subroutines. By contrast, as was indicated above, nothing like this would be ethically appropriate in a right-oriented setting.

Further Considerations Regarding Allocation

That aside, however, there does exist a certain parallel between right- and commodity-oriented healthcare institutions with regard to resources and allocation. In both cases, the administrators would have to keep in mind that their institutional resources are limited and that they are obliged to structure their organisational protocols accordingly. This means that in both cases there may be situations in which a decision would have to be made as to which patient would receive the indicated care when only one patient could be treated at a given time, and patients would have to be transferred to another institution or would have to be treated on a later occasion. Therefore, in both sorts of settings the AIs that the administrators would put in place to ensure that the relevant protocols would be followed would be designed to achieve consistent, effective and ethical implementation.

---

[17] American Society of Plastic Surgeons, *Plastic Surgery Statistics Report 2020*; accessed 23/10/2022 at https://www.plasticsurgery.org/documents/News/Statistics/2020/plastic-surgery-statistics-full-report-2020.pdf

[18] Boyd JB, Moon HK, Martin S, Mastrogiovanni DB. Plastic Surgery and the Malpractice Industry. *Plast Reconstr Surg.* 2021 Jan 1;147(1):239–247.

[19] Matej Mikulic, Alternative medicine industry revenue U.S. 2011–2021; accessed at https://www.statista.com/statistics/203972/alternative-medicine-revenue-growth/; "Complementary And Alternative Medicine Market Size, Share & Trends Analysis Report By Intervention (Botanicals, Mind Healing, Body Healing, External Energy, Sensory Healing), By Distribution Method, And Segment Forecasts, 2021–2028," accessed 25/10/2022 at https://www.grandviewresearch.com/industry-analysis/complementary-alternative-medicine-market

However, the protocols of such AIs in a commodity-oriented setting would have a different focus and contain different elements from those that were outlined with respect to a right-oriented institutional setting.

More specifically, and putting it first in general terms, the focus in a commodity-oriented as opposed to a right-oriented setting would be contractual in nature. That is to say, unless it was legally stipulated or otherwise required by such things as professional codes of ethics that the allocation protocols that hold in a right-oriented setting should also be followed, the initial and overall governing fact would be that the patients would in fact be clients. Consequently, the primary issue would be whether a patient was in possession of a service contract. A properly designed contract would include a priority ranking clause that applied in situations of allocation. If it did not, then the first come-first served rule would apply, just as it would in any other commercial setting. Consequently the AIs that would be designed to achieve consistent and effective implementation of the relevant allocation protocols in a commodity-oriented setting would find their focus in contractual matters and not in the fiduciary obligation towards the patients.

As was indicated before, exceptions to this might of course exist. Thus, it might be legally or professionally required that the type of allocation protocol that was outlined when discussing a right-oriented perspective should be followed. That, however, as was said, would be a matter of law or of professional codes. However, it would also mean that any AIs that an institutional administrator might use could not have a single unified operating system because different and conflicting types of considerations would then be relevant, and they would therefore have to be written into distinct subroutines. And while these might be located in one and the same computer, that computer could not have a single and unified operating system that dealt with the distinct logics. It would have to switch from one type of subroutine to another as the occasion demanded.

Concluding Considerations

If the preceding considerations are cogent, then the fundamental difference between a right- and a commodity-oriented perception of healthcare finds just as graphic a reflection at the administrative level as at the hands-on level, and it is mirrored in the design and the functioning of the AIs that might be used by the administrators of respective types of healthcare institutions.

However, as has been indicated, reality is rarely as cut-and-dried as purely theoretical considerations suggest, and the field of healthcare is no exception. For political, social and professional reasons—and even as a result of tradition—healthcare policies tend to be developed in a combination of right- and commodity-oriented lines. This means that, in reality, the factors that have to be considered relevant by an administrator with respect to the actual delivery of healthcare at the institutional level tend to form an intertwined series of elements.

In other words, institutional administrators in the field of healthcare tend to have to deal with logically incompatible premises because they have to manage healthcare institutions that are situated in a combined right-commodity oriented setting. Moreover, they also tend to have to juggle the demands of a fiduciary patient-oriented set of obligations that devolve on the healthcare professionals whom they employ with the demands of stockholder-oriented set of obligations who actually fund their institutions.

Among other things, this means that they have to develop some way of integrating the allocation criteria that are acceptable from a right-oriented perspective with those that are acceptable from a commodity-oriented perspective. Only if they can do that will the allocation protocols that they devise be functionally consistent and ethically as well as legally defensible. However, whether in fact they are able to do this is unclear because, as the preceding discussion has pointed out, that may involve reconciling inherently incompatible parameters. Therefore it is unclear how a single AI could assist them in that regard, because such an AI would have to integrate mutually conflicting subroutines into a single unified operating system, and that seems to be impossible.

Of course, as was also indicated, it is possible to design an AI that would switch from one legally and ethically distinct set of parameters to another as the occasion demands. Such a device would then be an AI that could handle these distinct perspectives. It would also be a device that a healthcare administrator could use and even incorporate into the framework of institutional decision-making protocols that they designed. Arguably, the working of such an AI would mirror the functioning of the human brain when it deals with logically incompatible matters in situations that involve cognitive dissonance. For, apparently, on that sort of occasion human awareness shifts from one of the dissonant items to another by using distinct neural pathways.[20]

That is to say, the human mind is able to handle mutually incompatible parameters in a single consciousness. For instance, it can handle distinct and mutually incompatible preferences, religious or legal tenets and apparently distinct analyses of physical reality such as relativistic and quantum-theoretical physics. Arguably, the fact that these distinct and apparently incompatible parameters can be subscribed to by one and the same individual and can be functionally present in one and the same consciousness at one and the same entails that the overall operating system of the human brain (which supposedly is the material basis of human awareness and is unitary despite involving distinctly functioning various parts) does so by means of what may be called shuttle thinking and switching between different parts of the brain.

---

[20] Marco Colosio, Anna Shestakova, Vadim V. Nikulin, Evgeny Blagovechtchenski and Vasily Klucharev, Neural Mechanisms of Cognitive Dissonance (Revised): An EEG *Study, Journal of Neuroscience* 17 May 2017, 37 (20) 5074–5083; Izuma K, Matsumoto M, Murayama K, Samejima K, Sadato N, Matsumoto K. Neural correlates of cognitive dissonance and choice-induced preference change. *Proc Natl Acad Sci U S A.* 2010 Dec 21;107(51):22014–9.

However, even if human consciousness is an entirely neurological and therefore material affair, the only way that this human ability could be reproduced in an AI is by organizing it in the way that was indicated above: Namely, by structuring it in materially distinct parts with different subroutines, where the different parts and their subroutines would then deal with right- as opposed to commodity-oriented issues and function from a utility as opposed to a deontological perspective as the occasion demanded.

However, while that is technically workable, it means that then the decision about which specific subroutine the AI should engage on a given occasion would ultimately be a matter of human decision making and not a matter of the AI's processing. In other words, it would mean that the decision according to which that happens would ultimately have to be handled by the institutional administrator or by some designated human being, since the AI would require some input that indicated which type of subroutine to employ. To be sure, an AI could assist an administrator in drafting the relevant sorts of protocols and their application. However, it would still mean that AIs could replace human administrators only in an operational and not in a decision-making role.

Needless to say, the problematic that has just been outlined could also be resolved by judicial means—that is to say, by stipulating in law how the administration of a healthcare institution should function, what sorts of protocols an institutional administrator should devise, and how the relevant AIs would be structured and function. However, that would not negate the previous considerations. Any AIs that would be used in this regard would still be tools. They would not be decision-makers in their own right.

Chapter 5
AIs and Healthcare Policy-Makers

Abstract The present chapter deals with ethical issues that arise for healthcare policy-makers who use AIs, how these are distinct from the issues that face hands-on healthcare professionals and institutional managers who use AIs, and the implications that this has for the design of the AIs that the decision-makers might use when carrying out their mandate.

Introduction

The preceding chapters discussed the use of AIs by healthcare professionals in the hands-on and the institutional administrative settings. It was suggested that although the kinds of ethical issues that they face overlap in many ways, how these play out differs because of the fundamental differences in the respective spheres of professional engagement. This, in turn—so it was suggested—entails that the AIs that are used in the one domain cannot simply be transferred for use in the other. Each of these domains requires not only its own AI usage protocols but also its own AI design.

Thus, ultimately it is the provisions that the policy-makers put in place that determine whether healthcare will be construed as a right or as a commodity. And while this has important implications at the hands-on as well as at the administrative level, the decision of how to structure the overall framework is something that neither hands-on healthcare professionals nor institutional administrators have to make. They simply have to adjust their practice in accordance with the policies that govern the setting in which they are situated, and they have to adjust the protocols of any AIs that they may use accordingly. It is the policy-makers who determine the nature of the principles that should be followed when being professionally engaged.

At the same time, the mandate of healthcare policy-makers is not confined to establishing the relevant policies. It also includes ensuring their consistent application. Consequently, they tend to establish designated agencies to this effect. The various subsidiary Departments and Centres of the National Institute of Health of

E.-H. W. Kluge, *The Ethics of Artificial Intelligence in Healthcare: From Hands-on Practice to Policy-making*, The International Library of Ethics, Law and Technology 25,
https://doi.org/10.1007/978-3-031-72837-2_5

the United States and of the National Authority for Health of France (*Haute Autorité de Santé*) are good examples of this. And while an analogous concern is involved in the institutional setting, the measures that institutional administrators put in place are different from the measures that have to be developed at the social policy level because the latter measures are overarching in nature and also have to deal with some different concerns.

To illustrate this by means of an example, while healthcare research is carried out by hands-on professionals and often takes place in an institutional setting, the rules that apply when such research is carried out are a matter of policy that is not determined at the hands-on or at the institutional but at the social policy level. Consequently, while an individual healthcare professional may decide to engage in healthcare related research, and while an individual institutional administrator may decide to allow such research to be carried out in the venue that they administer, what sort of research is permitted, as well as how it is structured and how it is carried out, is ultimately determined by the overarching research policies that have been put in place by the policy-makers. This is clearly illustrated by the fact that individual as well as institutional research protocols in the countries of the European Union have to be structured in accordance with the guidelines that are set at the level of the European Commission.[1] Similar considerations apply in North America[2,3] as well as in most other jurisdictions. Consequently, the AIs that the researchers might use in the conduct of their investigations have to be structured in accordance with the relevant provisions and, if they are appropriately designed, will effectively constrain the investigators to proceed accordingly.

As to the issue of policy formulation itself, it may be worth mentioning that while the perspective that the policy-makers adopt in formulating their policies—more specifically, whether they formulate them on a commodity- or a right-oriented basis—may be reflective of a general social consensus, that is not necessarily the case. Fundamental changes in the social perspective have been brought about by policy-makers on the basis of their own perspective. This is illustrated by the fact that how the nature of healthcare was originally viewed in Canada prior to the1940s[4] changed when, after much study and deliberation, the party in power adopted the view that healthcare is a right and introduced the *Canada Health Act*.[5] A similar

---

[1] European Commission, "Ethics for Researchers", accessed 30/11/2022 at https://ec.europa.eu/research/participants/data/ref/fp7/89888/ethics-for-researchers_en.pdf

[2] *Cf.* Govt. of Canada: Tri-Council Policy Statement: Ethical Conduct for Research Involving Humans—TCPS 2 (2022) accessed 01/06/224 at https://ethics.gc.ca/eng/policy-politique_tcps2-eptc2_2022.html

[3] *Cf.* Centers for Disease Control and Prevention, *Human Participant Protection in CDC Research,* accessed 23/11/2022 at https://www.cdc.gov/os/integrity/hrpo/index.htm; *Tri-Council Policy Statement,* available at http://www.pre.ethics.gc.ca/eng/policy-politique_tcps2-eptc2_2018.html

[4] MacDougall H. Into Thin Air: Making National Health Policy, 1939–45, *CBMH/BCHM* / Volume 26:22009 / p. 283–313; accessed 23/11/2022 at https://www.utpjournals.press/doi/pdf/10.3138/cbmh.26.2.283

[5] *Canada Health Act*, R.S.C., 1985, c. C-6; https://laws-lois.justice.gc.ca/eng/acts/c-6/page-1.html

phenomenon occurred when the various countries of Europe formed the European Union and its policy-makers, on the basis of their own understanding, established an overall health policy framework in the 1960s that construed healthcare as a right.[6] This, in turn, meant that the AIs that were used by the member states to regulate the structure of healthcare delivery in their own jurisdiction had to be amended accordingly. Consequently, as will be discussed in greater detail in what follows, this required that the relevant AIs had to have access to different data and had to use different kinds of algorithms.

Moreover, the policy-makers' perspective on whether healthcare is a right or a commodity is not the only things that influences their approach to healthcare policy-making. So is their understanding of what constitutes an appropriate ethical framework under which healthcare policies should be formulated: More specifically, whether it should be from a deontological or a utilitarian perspective. This also determines the structure and functioning of any AIs that they might use when engaged in their policy formulation.

To illustrate how this is the case, it may be appropriate to indicate how the construal of healthcare as a commodity and the construal of healthcare as a right each results in different policies when different ethical approaches are adopted by the policy-makers, and relate this to the design of the AIs that may be used in this connection.

General Considerations

Healthcare Construed as a Commodity

When healthcare is construed as a commodity—and ignoring for the moment the ethical perspective that the policy-makers might adopt—then the major considerations that would be relevant in the policy-makers' decision-making would be considerations that find their focus in the economic well-being of their society. Consequently, if they act in keeping within the nature of their mandate, their central concern would essentially be similar to what would be of concern to other social policy-makers with respect to other types of commodities. Thus, customer safety would be a major issue, as would be economic growth, innovation, veracity in advertising, the nature of contracts, etc.[7] Moreover, unless it was reasonable to assume that putting in place a publicly funded healthcare system was indicated for the efficient functioning of society itself, the policy-makers would not establish

---

[6] HH Yildirim, et al., European Union Health Policy: A Short Overview, *Journal of Health Systems and Policies*, Vol.1, No:3, 2019.

[7] Cf. US Department of Commerce, "Strategic Goals", available at https://www.commerce.gov/about

such a system. Healthcare would then be just like any other commodity, and there would be no overriding societal obligation to provide it.

Ethical Considerations

Furthermore, while in many ways the issues that would concern the policy-makers if construed healthcare as commodity and adopted a utilitarian perspective would be similar to those with which they would deal if they adopted a deontological perspective, the reasoning that would they would use in formulating their policies would be different even though in each case they considered healthcare to be a commodity. Moreover, while some technical aspects of the AIs that they might use when formulating the relevant policies might also be similar, how the AIs would be used—which would include such things as the data that would be fed into the AIs—as well as how they would be designed would be different.

More specifically if, while construing healthcare as a commodity, they adopted a utilitarian perspective, then their considerations would focus on whether imposing a particular policy—for instance, regarding patient safety or professional licensing or practice requirements, etc.—would result in the greatest amount of good for the greatest number of people in their society. Therefore, resource allocation in the sense that was discussed in the preceding chapter with respect to a deontological framework of allocation would not enter their considerations. Nor would this be reflected in the algorithmic design of the AIs that the policy-makers might use since, all other things being equal, there would be no greater overarching societal mandate to provide healthcare than there would be for any other commodity. Consequently, such factors would not enter into the algorithms of the AIs that they might use. Moreover, calculative measures such as quality adjusted life years (QUALYs)[8] and disability adjusted life years (DALYs)[9] would be appropriate and could well be integrated into the subroutines of any AIs that they might use when formulating their policies.

By contrast, if the policy-makers adopted a deontological ethics and still viewed healthcare as a commodity, then could not consistently make use of such measures as QALYs or DALYs in their considerations. Instead, their considerations would have to be in accordance with ethical Principles such as Autonomy and Equality. Moreover, they would formulate their policies by considering what effect their policies would have on individual persons, and the effect that they would have on "the greatest number of people" would at best be a secondary matter of concern.

[8] Prieto L, Sacristán JA. (2023). Problems and solutions in calculating quality-adjusted life years (QALYs). *Health Qual Life Outcomes*. 19;1:80.

[9] Murray CJL. (1994). Quantifying the burden of disease: the technical basis for disability-adjusted life years. *HO Bulletin OMS*. 72: 427–445.

Therefore their considerations would essentially be similar to considerations that applied in deontologically oriented business ethics.[10]

With due alteration of detail, these factors would be reflected in the design of any AIs that they might use in their deliberations.

Healthcare Construed as a Right

On the other hand, if the policy-makers construed healthcare as a right, then they would be obliged to put in place a public healthcare system.[11] And here the fact that no society has an infinite amount of resources would have important implications not only with respect to the types of healthcare that the society for which they are responsible would provide but also with respect to how it would provide them. For instance, whether it would provide it in localised or in distributed settings, by means of personal contact between patients and professionals or by mediated means such as eHealth and telemedicine, etc.

Ethical Considerations

Moreover, even if the policy-makers agreed that healthcare should be construed as a right, the policies that they put in place would also be affected by the ethical perspective that they adopted.

For instance, if the policy-makers adopted a deontological ethical approach, they might well decide that interventions of a purely cosmetic nature would not fall under the rubric of "right-to-healthcare" even if such interventions were medical in nature, and that an exception to this would only be granted if an affected person could not only show that they suffered from a disfigurement that could be ameliorated only through an appropriate healthcare intervention, but also that this disfigurement met the equity criterion: More specifically, that it interfered with the individual's ability to take advantage in an equitable fashion of the opportunities that the society offered to its members. Once more, this would require fundamental provisions in the subroutines of any AIs that they might use in making their decisions.

By contrast, if the policy-makers followed a utilitarian ethical approach when executing their right-oriented mandate, then they would be obliged not to structure the overall policy framework that they designed in keeping with the deontological principles of Equality, Autonomy, etc., unless purely legal reasons such as the constitution of their society constrained them to do otherwise. Instead, they would

[10] Micewski ER, Troy C. (2007). Business Ethics—Deontologically Revisited. *J Bus Ethics* 72, 17–25.

[11] Cf. Kluge E-H. W. (2022). *The Right to Healthcare: Ethical Considerations*. Chapter 3.

structure it in keeping with the Principle of Utility and its derivatives. In other words, their major concern would then be how to achieve the greatest amount of good for the greatest number of people. This, in turn, would mean that the AIs that they might use in their deliberations would involve the very calculative parameters that could not be used if the policies were planned according to a deontological perspective. Therefore, they could then include such measures as QALYs and DALYs, etc. In this way the whole ethical framework within which they as policy-makers moved would shift, and so would the structure and design of the AIs that might assist them in performing their task.

Moreover, with respect to making such a decision—and this applies both with respect to a right and a commodity focused perspective, as well as with respect to adopting either a deontological or a utilitarian orientation—the only role that AIs would here play would be to present the implications of adopting a particular perspective insofar as these implications could be worked out by algorithms that are calculative in nature.

In other words, while the health policy decision-makers could use AIs in making their decision, the AIs themselves could not answer the question which perspective they should adopt. They could only provide the alternative calculations on the basis of which the policy-makers would then make their decisions. Therefore, which approach the policy-makers should adopt would not be a matter with respect to which the calculations of the AIs could be conclusive. The ultimate decision would involve an interplay between the personal convictions of the policy-makers and the societal perspective in which they operated.

Resource Allocation

At this point, it may be useful to consider in somewhat greater detail how the matter of resource allocation would enter the policy-makers' deliberations, for the issue would present itself in different ways depending on whether they adopted a right-oriented or a commodity-oriented perspective, and whether they adopted a deontological or a utilitarian ethics. It would also impose different requirements on the design of any AIs that they might use in their deliberations.

Healthcare as a Right

To put it first in general terms, if the policy-makers considered health care to be a right that belonged to every qualified member of their society, then the question of what should be included in the healthcare system that they designed would be determined on the basis of how the ethics of healthcare delivery would play out within the framework of healthcare construed as a social obligation.

Also, if healthcare was considered to be a right—and ignoring for the moment the distinction between a deontological and a utilitarian approach—the policy-makers' initial task would be to determine who should have access to what would effectively be a public healthcare system.

However, that decision has much deeper roots than would appear at first glance. For, ultimately, it is a decision that is based on the judgement whether the right to social services belonged to all persons who are present in that society irrespective of whether they are permanent residents or even citizens, or whether such a right belonged only permanent residents and citizens. Thus, if it belonged to all persons, then anyone who presented in the web of social interactions would be entitled to take advantage of the healthcare system that the society provided. On the other hand, if it only belonged to individuals who met residence or citizenship requirements, then only these would be entitled to whatever healthcare services were put in place. The health policy decision-makers could then use AIs to calculate the costs that would be involved on these different assumption and what implications this would have for the overall resource pool that was available to the society for fulfilling all of its obligations.

Once More, the Relevance of Ethics

Moreover, the matter of ethics enters once again. For, even if the policy-makers construed healthcare as a right, which types of healthcare would be provided as a matter of policy would depend on whether they viewed healthcare from a deontological or from a utilitarian perspective. And this also would have important implications for the design of AIs that they might use in their deliberations. Thus, if they adopted a deontological perspective, then the AIs' calculations would involve action gradients that would be grounded in deontological Principles. On the other hand,, if the policy-makers adopted a utilitarian perspective, then deontological Principles such as the Principle of Equality would play only a minor role—if indeed they played any role at all. What would then play a major role would be the Principle of Utility, and the AIs would calculate what kinds of healthcare modalities would have the best possible outcomes for the greatest number of people.

A Utilitarian Perspective

More specifically, the implications for the design of a right-oriented public healthcare system from a utilitarian perspective can be put in several overarching points: First and most obviously, only the modalities that addressed the health conditions of the majority of individuals would be funded. Consequently, the conditions that affected only a minority of individuals would receive funding only after the needs of the majority had been met.

At the same time, however, it would also mean that even the health conditions that affected the majority of individuals would only conditionally be funded. More specifically, they would be funded only if these conditions detracted from what could be identified as constituting "the amount of good" with respect to their sufferers. Therefore, if that good was understood in terms of health status—an issue which, as will be discussed in a moment, is not without its own difficulties—any condition that did not greatly affect the majority's health status or their ability to contribute to the resource pool of society would not be included in the design of the healthcare system. The common cold would be an example in this regard, as would be simple headaches o asymptomatic flatulence. After all, the reason that the matter of the majority's ability to contribute to the society's resource pool would enter the picture is that a society cannot institute a healthcare system without sufficient resources—not even from a utilitarian perspective.

However, this means that the policy-makers and the AIs that they might use in their deliberations would require appropriate data for their decision-making: data that correlated health conditions with health status and productivity, as well as data that identified the number of individuals who were affected; because it is only such data that would allow a correlation between healthcare issues and economic performance. In other words, at this point the issues of healthcare and economics would intersect in the matter of policy design—which in turn introduces the question whether such a healthcare system could realistically be designed independently of the design of an economic system, as well as the question whether the initial macro-allocation of the society's resources into distinct types of social services would have to be an integrated procedure that requires iterative adjustment as conditions changed. If that is the case—and it certainly seems to be—then any AIs that the policy-makers might use when making their decisions would also have to be able to bridge these distinct domains.

Subsidiary Considerations

This, however, brings up another issue: namely, the matter of privacy. The reason for this is that privacy considerations arise both in the field of economics and the field of healthcare. However, the conditions that are considered ethically relevant in the economic[12] and the healthcare spheres[13] differ. Consequently, the ability to correlate the distinctive ethical requirements of these domains would be necessary a consideration in the design of any AIs that might be used in the formulation of health policies.

A way of doing so, of course, would be to accept that the functioning of AIs does not require identified data: that only the data themselves are important. However, it would still mean that an appropriate provision for developing and providing these data would have to be contained in the overall funding schema of the healthcare

[12] Acquisti A, Taylor C and Wagman L. The Economics of Privacy, *Journal of Economic Literature* 54: 2, June 2016 (pp. 442–92).

[13] WHO, *The protection of personal data in health information systems—principles and processes for public health,* https://apps.who.int/iris/bitstream/handle/10665/341374/WHO-EURO-2021-1994-41749-57154-eng.pdf?sequence=1&isAllowed=y

policy-makers. In other words, a separate funding item with this provision in mind would have to be included in the overall healthcare resource allocation schema that the policy-makers designed. Moreover, it would mean that any AIs that they used in the conduct of their deliberations would have to have the requisite data access and the requisite data handling provisions.

Moreover, the nature of what constitutes the good would also become an issue for utility-oriented policy decision makers even when they construed healthcare as a right; and this again has implications for the design of any AIs that they might use in their deliberations.

More specifically, what counts as "the good" depends on the conceptual perspective of the society in question. So, for instance, the conceptual framework of a society that sees a healthy mode of living as not necessarily involving longevity and freedom from disease but as living in a traditionally respectful way and in an integrated and symbiotic manner with the rest of nature,[14] or the conceptual framework of a society that combines magical with traditional and purely pragmatic considerations[15] would here be illustrative examples. However, when a society comes to adopt a purely scientific perspective about the nature of reality, then its conceptual framework changes and this has ethical implications. More specifically, the society then becomes obliged to engage in research in order to determine whether the application of the scientific approach to the delivery of healthcare showed that the traditional or magical healthcare modalities still maximised "the good", or whether new types of healthcare interventions should be substituted for the old: In other words, it becomes obligated to determine whether the old perspective should be replaced by or be modified in keeping with the new scientifically grounded understanding of the nature of disease and of disease process, etc. or whether the old approach should continue to should continue to determine the identity of "the good". This, however, can be determined only on the basis of research and experimentation. Consequently, while the policy-makers would likely use AIs in the conduct of that research, such AIs would have to be designed in a suitable manner.

Likewise, only properly conducted research would show whether attaining a higher standard of "the good" for a particular healthcare modality was in fact likely or even possible. Only in this way would it be possible to determine whether the effectiveness standards that obtained at a given point of time could be improved. For, the ability to maximise the good is dependent on the scientific and industrial conditions that obtain, and therefore is not a static matter. This, in turn, means that a policy that aimed to provide the greatest amount of good for the greatest number of people would have to contain relevant temporal and research parameters. Once more, therefore, the AIs that might be used by the policy-makers in this regard would have to be appropriately designed and would have to contain access routines to the relevant data bases.

---

[14] First Nations Health Authority, *First Nations Perspective on Health and Wellness*, https://www.fnha.ca/wellness/wellness-for-first-nations/first-nations-perspective-on-health-and-wellness

[15] Lam MT, Lam HR, Chawla L. Application of magic in healthcare: A scoping review. *Complement Ther Clin Pract.* 2017 Feb;26:5–11

Moreover, even when a utilitarian ethics is adopted on these matters, what counts as "the good" is itself a relative matter. Thus, ideal as opposed to hedonistic utilitarianism differ on this point.[16] Healthcare policy-makers would therefore have to decide at the outset which interpretation of "the good" to adopt and how to integrate it into the design of any AIs that they might use.

At the same time, as was pointed out before, even when there is agreement on that issue, and even ignoring the issue of scientific sophistication that was mentioned above, another fact that was hinted at but that so far has not received sufficient attention is that irrespective of how the qualitative nature of "the good" is understood, what constitutes the fulfilment of that good is not a static matter. Thus, maximizing "the good" at one point in time is not the same as maximising it at another time. Conditions change in the demographic,[17] genetic[18] as well as epidemiological[19] and similar parameters that affect health status, and all of these are functionally interrelated. This in turn means that when a healthcare policy is intended to maximize the health status of the greatest number of people, the quantitative expression of this would shift as these conditions changed. Calculating this, however, requires that the AIs that might be used in this connection have access not only to population statistics in terms of pure population numbers but also with respect to demographic data that include genetic and epidemiological factors.

And this, once more, also has resource implications, because the resources that would be necessary to underwrite modalities that would lead to the best possible health status at one point in time may differ from the resources that would be required to fund the modalities that would be necessary to achieve this at another point in time. Consequently, the calculative algorithms of any AIs that the policy-makers might use in their decision making would also have to reflect this. In order to achieve this, a feed-back routine that would correlate resource availability, healthcare modality effectiveness and health status would have to be built into the design of the AIs.

That, of course, would be impossible unless the AIs had calculative routines that allowed them to assess the effectiveness of the various healthcare modalities. Moreover, this effectiveness would have to be calculated not on an individual-incidence basis but on a statistical basis. After all—as was previously pointed out—this is required when calculating the greatest amount of good for the greatest number of people. Therefore except for the case where the majority of members of society were identical in their reactions to a given healthcare modality—which is highly

[16] *Stanford Encyclopedia of Philosophy* "History of Utilitarianism, https://plato.stanford.edu/entries/utilitarianism-history/

[17] NIH, National Institute on Aging, "Global Aging", https://www.nia.nih.gov/research/dbsr/global-aging

[18] NIH: National Human Genome Research Institute, "Genetic Drift," https://www.genome.gov/genetics-glossary/Genetic-Drift

[19] Santosa A, Wall S, Fottrell E, Högberg U, Byass P. (2014). "The development and experience of epidemiological transition theory over four decades: a systematic review". *Global Health Action*. 7: 23574.

unlikely—this means that the effectiveness of a given healthcare modality would identify a range and not a specific numerical figure. One more, this means that the AIs that might be used would have to be able to access the relevant data bases in a continuing manner.

A Deontological Perspective

On the other hand, if the policy-makers accepted that healthcare should be considered a right but adopted a deontological perspective, then their primary considerations would centre in making sure that the services that would be included in the healthcare system that they designed would be in keeping with the Principle of Equality. That, in turn, would require that any AIs that would be used by the policy-makers would have data that could establish whether providing a particular healthcare modality would (likely) potentiate its recipients' ability to take advantage in an equitable manner of the opportunities that the society offered to its members. And that, in turn, would not only mean that such AIs would have to be able to access the health records of each individual person, but that they would also have to contain subroutines that could compare the health status prior to and subsequent to receiving an intervention in the relevant manner.

In other words, the issue of identified vs. de-identified data would here arise as well. However, unlike the situation that would exist when a right-oriented approach to healthcare is construed from a utilitarian perspective, it could not be resolved by having the relevant AIs simply deal with numbers that statistically determined whether the majority of individuals were provided with healthcare that resulted in the greatest good for the greatest number. Nor could any AIs that are used in this connection be able to calculate using such measures as QALYs or DALYs. For, as was indicated in Chap. 1,[20] using such measures would not indicate whether the relevant policies would provide each individual with access to healthcare in an equitable manner. The Principle of Equality would be satisfied only if the relevant policy ensured that every qualified individual was provided with the relevant services in an appropriate manner. As will be indicated later,[21] the AIs that could be used in a deontological setting would therefore have to use entirely different methods of calculation. Moreover, they would require access to the health records of each individual who would be affected by the relevant allocation policy.

Therefore, while neither a right-oriented system that was designed from a deontological nor one that was designed from a utilitarian perspective could ignore outcome considerations, how these would be involved in the calculations of their respective AIs would be fundamentally different. If the system was designed from a utilitarian perspective, the aggregate outcomes would be determining, and the question whether the preponderance of individuals would be able to contribute in a

[20] Cf. pp. 10 *et pass.*

[21] Cf. p. 101, *infra.*

positive manner to the functioning of society would be the important consideration. By contrast, while outcomes would also be considered relevant if the system was designed from a deontological perspective, the deciding issue would be whether each individual who received the relevant services responded to the interventions by being able participate in an equitable manner in the opportunities that society offers. Whether they would be able to contribute to the functioning of society as a functioning enterprise would be a secondary consideration.

The Issue of Healthcare Itself

Another matter that has essentially been ignored in the preceding considerations but that is a crucial determinant in the design of a healthcare system at the policy level is how to understand the notion of healthcare itself. For, ultimately, it is this that determines what should be included under the rubric of "healthcare" and how the relevant AIs should be designed.

As was pointed out in Chap. 1,[22] the problem here is that there are a variety of ways in which the notion of health itself can be understood.[23,24,25,26,27,28,29] Therefore this would affect what modalities would be included in the array of interventions to be provided by a publicly funded healthcare system irrespective of whether it was designed from a utilitarian perspective. Consequently, it would affect also the design of any AIs that the policy-makers might use in their deliberations. For, it would determine what kinds of data the AIs would have to be able to access and even the value gradients on the basis of which they would rank the modalities relative to each other.

However, even that would be insufficient if the policy decision makers were to be able to use AIs in their decision making. Thus, for political reasons—and this is based on the assumption that the policy-makers would be democratically elected rather than would autocratically assume their position—the policy-makers would also be responsive to the values that were shared by the members of their respective societies. Therefore, the value gradients that would have to be incorporated into the

---

22 Cf. pp., 3 ff.

23 Engelhardt T, "The Concepts of Health and Disease," in *Evaluation and Explanation in the Biological Sciences* Engelhardt and Spicker (eds.), Dordrecht, Reidel, 1975, 125–141.

24 Nordenfeldt L, *On the Nature of Health: An Action-Theoretic Approach* Dordrecht: Reidel, 1987).

25 Whitbeck C, "A Theory of health" in Engelhardt and McCartney, eds. *Concepts of health and Disease: Interdisciplinary Perspectives.* Reading, Mass: Addison-Wesley, 1982; 611–626

26 Boorse C, "Health as a theoretical Concept," *Philosophy of Science* 44 (1977) 542–573

27 Bircher J, Hahn EG. Understanding the nature of health: New perspectives for medicine and public health. Improved wellbeing at lower costs. F1000Res. 2016 Feb 12;5:167

28 Jadad AR, O'Grady L: How should health be defined? *BMJ*.2008: 337, a2900.

29 Kelman S. (1975). The social nature of the definition problem in health. *Int J Health Serv.* 5(4):625–642.

design of the AIs that they might use would not be the individual values of the policy-makers themselves but the values that were shared by the individuals who put the policy-makers into office and who imbued them with decision-making power. This, in turn, means that these value-gradients would have to be responsive to changes in the population's values. Consequently, the design and functioning of any such AI would also have to include access to public value perceptions, and that would never be an absolute and finished matter either, but would be subject to constant re-evaluation and revision. Quite aside from the issue of how this could be achieved in the design and functioning of the relevant AIs, the issue of expenditure once more arises: i.e., the expenses that would be associated with this with respect to the design and functioning of the relevant AIs.

A Deontological Approach

It goes almost without saying that if the policy-makers adopted a deontological perspective and viewed healthcare as a right, then they would here base their decisions on the deontological Principles that were outlined in the initial chapter: namely, Autonomy, Equality, Impossibility, Beneficence and Nonmaleficence.

More specifically, the Principle of Autonomy would shape the framework that they promulgated by containing as a fundamental determinant that, all other things being equal, access to the healthcare system that they instituted was on a voluntary basis. The only exception that would be recognised in the policy framework would be when someone's refusal to access the system or otherwise to deal appropriately with a health condition would impair the health of other members of society. This would be ethically justified by the conditioning clause of the Principle of Autonomy itself, as well by the Principle of Nonmaleficence.

As to the Principle of Equality, its import would affect not only the types of healthcare services that would be included in the healthcare system, but also that access to the services would be an equitable manner.

With respect to the first of these conditions, the fact that no society has an unlimited amount of resources would again entail that the resources that would be available for healthcare would also be limited. Since each healthcare modality would constitute an expenditure item and therefore the resources that would be expended in one area would not be available for any other, this means that including a particular treatment modality in the healthcare system would have to be justifiable in the sense that it would (likely) potentiate its recipient's ability to take advantage in an equitable manner of the opportunities that the society offered to its members, but that this would not affect the other healthcare relevant modalities that also satisfied the Principle of Equality.

At first glance it might seem the policy-makers ought then to construe their mandate as being that of designing a healthcare system that was likely to ensure that each member of society had the same health status. And if that was the case, then the AIs that they might use in their policy design would evaluate not only the health status of each individual person but whether the relevant provisions met this equality condition.

However, to construe their mandate thus would be to ignore the implications of the Principle of Impossibility on the Principle of Equality. More specifically, it would be to ignore the fact that it is impossible to guarantee that every person has the same or equal health status. So, for instance, diseases such as Angelman syndrome,[30] autosomal dominant polycystic kidney disease[31] and Huntington's disease[32] are in principle incurable, as are such conditions as muscular dystrophy,[33] haemophilia and narcolepsy,[34] to name but a few conditions. No matter what healthcare modalities may be employed to deal with them, short of including effective genetic modification techniques in the domain of healthcare modalities, it will be impossible to ensure that those who suffer from them will ever have an equal health status as those who do not suffer from them. In fact, short of making all individuals genetically identical as well as identical in lifestyle—and, arguably, that would run head-on into the Principle of Autonomy[35]—equality in health status cannot ever be achieved. Consequently, to construe the task of deontologically oriented healthcare policy decision-makers to be that of designing a healthcare system that achieves the same health status for every individual would be an error.

It would also be to misinterpret the implications of the Principle of Equality—at least for the field of healthcare. The Principle of Equality does not require that healthcare be directed towards achieving sameness in outcome because, as was just pointed out, that is impossible, and an impossibility cannot constitute an obligation. As already Kant pointed out, ethically speaking "ought" implies "can".[36] This even holds in the juridical sphere, as was formally acknowledged in the English case of *Taylor v. Caldwell*[37] and in fact goes back to old Roman law.[38] It is also a principle in legal systems that do not fall into the European tradition, as for instance the legal

---

30 Maranga C, Fernandes TG, Bekman E, da Rocha ST. Angelman syndrome: a journey through the brain. *FEBS J.* 2020 Jun;287(11):2154–2175

31 Colbert GB, Elrggal ME, Gaur L, Lerma EV. Update and review of adult polycystic kidney disease. Dis Mon. 2020 May;66(5):100887. https://doi.org/10.1016/j.disamonth.2019.100887. Epub 2019 Sep 30. PMID: 31582186.

32 Conditions that cannot be cured, https://www.qld.gov.au/health/support/end-of-life/care/conditions#:~:text=cancer,neurone%20disease%20and%20multiple%20sclerosis

33 Arora H. Duchenne muscular dystrophy: Still an incurable disease. *Neurol India.* 2019 May-Jun;67(3):717–723.

34 NIH, National Institute of Neurological Disorders and Stroke, "Narcolepsy", https://www.nhs.uk/conditions/narcolepsy/treatment/#:~:text=There's%20no%20specific%20cure%20for,impact%20on%20your%20daily%20life

35 Hague R, *Autonomy and Identity: The Politics of Who We Are* (Routledge, 2011); Marina A.L. Oshana, "Autonomy and Self-Identity," in John Christman and Joel Anderson, eds. *Autonomy and Challenges to Liberalism* (Cambridge University Press, 2009).

36 Kant, Immanuel. *Critique of Pure Reason*. A548/B576. p. 473.

37 Taylor v. Caldwell; 3 B.&S. 826, 122 *Eng. Rep.* 309 (K.B. 1863).

38 Justinian, *Digest* https://www.thelatinlibrary.com/justinian/digest50.shtml

system of the People's Republic of China.[39] Instead, the Principle requires that an equitable effort be expended with regard to the individuals in question. This means that the Principle would be satisfied if the modalities that the policy-makers put in place ensured that an equitable amount of effort was expended on behalf of each individual with respect to healthcare.

The question, of course, is how that could be measured without reverting back to the criterion of equal outcome: i.e., without insisting that this would be the case only if the healthcare needs of each individual were successfully addressed and that the health status of each individual was the same. For if that was the only way then, as was said, deontologically oriented policy-makers would be faced with an impossible task and would essentially have to abandon their deontologically oriented approach.

Arguably, a way of avoiding this would be to measure the effort in terms of the quantity of funds that are expended in providing the relevant treatment modalities. If that was accepted, then this would not necessarily mean that the relevant modalities had to have the same health status outcome because, even though equal amounts of funds might be expended, the outcomes might not be the same due to the natures of the individuals and of the health conditions themselves.

However, this way of measuring equitability of effort would also be unacceptable. Not all health conditions affect their sufferers in the same way, and the kinds of interventions that are appropriate for dealing with them varies in the nature, intensity and frequency of the application of the relevant modalities, and this varies as costs change. Arguing that expending the same amount of funds for all health conditions would satisfy the Principle of Equality would be to ignore this, and therefore would be to ignore an important sub-clause of the Principle itself that was pointed out in Chap. 1—namely, that ethically relevant differences in the nature or status of the person in question mandate a difference in treatment.

That is to say, the differences in the natures of the health conditions from which different individuals suffer require that they be addressed by using different types of healthcare modalities if the fiduciary obligation of healthcare providers is to be satisfied. "One size fits all" is not a maxim that applies in healthcare. Moreover, healthcare modalities require different amounts of funds at different times for their application. Therefore, since this difference is tied to the difference of the relevant health conditions, and since the subsidiary clause of the Principle of Equality the latter recognizes ethically relevant differences, it follows that sameness of funds-expended cannot be used as an ethically appropriate measure.

The central point here is that both the suggestions of sameness-of outcome and of sameness-of-funding ignore the fact that the Principle of Equality is not a principle of sameness in a mathematical sense but is a principle of equity. Therefore, it would be satisfied if and only if the policy directives that healthcare policy-makers put in place stipulate that individuals who suffer from different health conditions be

---

[39] Wang J. "The Rule of Law in China: A Realistic View of the Jurisprudence, the Impact of the WTO, and the Prospects for Future Development." *Singapore Journal of Legal Studies*, 2004, pp. 347–89.

treated equitably. And in the present context, this means that in order to satisfy the Principle of Equality a healthcare policy would have to mandate that an equitable amount of effort be expended on dealing with different healthcare conditions and with the individuals who suffer from them, since they suffer from different health conditions and suffer in different ways.

Act-Effectiveness Coefficients

This, however, points to a way in which the policies that are put in place by the policy-makers who adopt a rights-approach to healthcare as well as a deontological perspective may be constructed. That is to say, the differences between different kinds of modalities consist not only in the methodologies, tools and frequencies with which they are employed but also in their relative effectiveness. For instance, by and large the kinds of treatment that are appropriate for dealing with a broken leg tend to be more effective than the types of treatment that are appropriate for dealing with such mental illnesses as schizophrenia, Alzheimer's disease or anorexia nervosa.

At first glance, it seems that adopting this sort of measure would be to revert to the outcome-criterion criterion after all, and therefore would be subject to the same objection. After all, the effectiveness of an intervention is measured in terms of its outcome.

While in a sense that is correct, it is correct in a sense only, and to argue thus would be to ignore the fact that the outcomes in question are different in nature. The outcome that was at issue in the initial discussion was the overall health status of the individuals in question, whereas the outcome in this instance is the effectiveness of the intervention with respect to the particular condition and the particular individuals to which it is applied.

Thus, the very same type of treatment may be differently effective for different individuals even when it is employed using the same methodology and the same types of tools, depending on the individuals' respective genetic make-up, co-morbidities, etc. This, in turn, means that the effectiveness of a given intervention is not appropriately represented by a single figure but by a range that spans the degree of effectiveness that the particular intervention has for different individuals who suffer from the same condition. If this range is called its act-effectiveness coefficient, then it may be used to measure the effort that is expended in dealing with the conditions. For, then the equity criterion with respect to expenditure of effort would be satisfied if the effectiveness of an actual intervention falls within the act-effectiveness coefficient of other types of intervention in question without at the same time requiring that the overall health status that results from its application be the same.

This, of course, requires that it be possible to measure the amount of effort that is expended in a given situation without reverting to the criterion of funding or health status, and thereby essentially eventuating in an ethical deadlock. However, if "effort" is understood as "a vigorous or determined attempt" as the Oxford

English Dictionary has it,[40] or as "a serious attempt" as it is put in the Merriam-Webster Dictionary,[41] then the amount or degree of effort that is expended in a given situation can be determined independently of purely financial considerations by considering the dedication or commitment with which the condition in question is addressed.

Arguably, this dedication or commitment can be measured by asking whether the provisions that are put in place to deal with the relevant conditions satisfy a particular framework requirement: More specifically, that they enable the individuals who receive those intervention to participate in an equitable fashion in the opportunities that society offers to its members as a result of receiving those interventions whereas otherwise that would not be the case. Moreover, this can be extended as an overarching condition to all kinds of conditions and interventions. Therefore, using the terminology that was introduced little while ago, it means that if there is a difference in the act-effectiveness coefficient of a particular type of intervention compared to others, and if that act-effectiveness coefficient is lower than that of the other types of interventions, then an equitable amount of effort would be expended if and only if "a vigorous or determined attempt" was made to bring the act-effectiveness coefficient of the one that has the lower range up to the level that was characteristic of the other types of interventions.

More specifically still, when the act-effectiveness coefficients of the distinct healthcare modalities that are offered by a society are compared with each other, it turns out that these coefficients converge towards a median. Therefore, when the act-effectiveness coefficient of a healthcare modality hits that median—in other words, when it has the same range as the median of the act-effectiveness of the average healthcare modalities that are provided for by the society—then, irrespective of the amount of funds that are expended in that connection, the effort that is expended when dealing with the condition is equitable. On the other hand, if the act-effectiveness coefficient does not hit the median, then an equitable healthcare policy would mandate that appropriate steps be taken to correct the situation. It then takes into account the differences between the relevant health conditions, the severity with which the affected individuals suffer from them, the individuals' co-morbidities and other relevant factors.

To be sure, this does not mean that funding drops out of the picture. However, it means that funding itself is not the determinant but the act-effectiveness coefficient. Therefore if the act-effectiveness coefficient of a given healthcare modality falls below this average value—or if in fact there is no appropriate healthcare modality at all, as would be the case with some currently incurable or untreatable conditions—then deontologically oriented healthcare policy-makers who construed healthcare as a right would adjust healthcare funding in such a way that the coefficient was brought up to the median. This may involve either increased funding for the area in

[40] *Oxford English Dictionary,* https://www.google.com/search?q=effort&rlz=1C1AFAB_enCA446CA446&oq=effort&aqs=chrome..69i57j0i433i512j46i433i512j46i175i199i512j0i433i512j0i512j46i512j0i512j0i433i512j0i512.2110j1j15&sourceid=chrome&ie=UTF-8

[41] *Merriam-Webster Dictionary*, https://www.merriam-webster.com/dictionary/effort

question or it may involve allocating resources to designated research in that particular area. In this way effort, effectiveness and research could be integrated into equitable healthcare policy-making and would satisfy the Principle of Equality.

In order to this approach to be workable, however, and in order to be able to use AIs in their deliberation, several kinds of data would be necessary, and these data would have to be available to any AIs that the policy-makers might use. Thus, data would be required about the array of health issues that are encountered by the members of the society for whom the policy-makers are designing a healthcare system. Data would also be required about which of these health care conditions naturally interfere with their sufferers' ability to participate in an equitable manner in the opportunities that exist within that society for—as was indicted previously—not all health conditions have that effect. A third datum that would be required would be the proportion of individuals who are thus affected, and a fourth datum would be an account of the healthcare modalities that are available within that society for dealing with these various health conditions. A fifth datum would be an account of the healthcare conditions that meet the condition that falls under the second rubric—i.e., of conditions that naturally interfere with their sufferers' ability to participate in an equitable fashion in the opportunities that society offers—but where no healthcare modalities for ameliorating or otherwise addressing the issue are available, where the absence of an available treatment modality would count as having a zero act-effectiveness coefficient. A final datum would be the actual number of individuals who suffer from the types of health conditions that fall under the various rubrics that were previously mentioned. Using these data, the AI that the policy-makers might use could then determine how the treatment modalities ranked relative to each other with respect their act-effectiveness coefficients and whether steps would be necessary to bring them to the same range. This could then be integrated by the policy-makers into a policy that either allocated more funds for relevant treatments, for research or for palliative care.

Further Considerations

A general point of significance and one that so far has not even been mentioned is that even if different societies subscribed to the perspective that healthcare is a right, the preceding considerations do not entail that in order to fulfil the equity criterion—in other words, in order to satisfy the Principle of Equality—the healthcare policy-makers of all similarly oriented societies must use the very same act-effectiveness coefficients when designing their respective healthcare systems. In other words, it does not follow that the act-effectiveness coefficiency of a given intervention constitutes a universal measure and that different societies may not differ in the value that they attach to this measure. The reason for this is that societies are distinct entities, and that the resources that they have available for providing the services that they ethically obliged to provide if they construe healthcare as a right will differ. Therefore, the preceding condition of equitable treatment could be met even though the act-effectiveness coefficients of the very same healthcare modalities differed from society to society. The only time when this would not be the case would be of the society in question was a global society, because then one would be dealing with a single social and resource system.

Moreover, another matter that so far has been ignored but that plays a significant role in healthcare decision-making is the fact that in healthcare there are very few if any certainties,[42,43,44] and that just like the decisions that are made by professionals at the institutional and the hands-on level, the decisions that healthcare policy-makers make are not only probabilistic[45] but also valuational in nature. Since the values as well as the healthcare needs of society shift over time, the division of healthcare funds into research and healthcare delivery modalities requires constant attention. Once more, this would mean that it would have to be accommodated in the design of the relevant AIs and would mean that their actual calculations would be a constantly shifting affair.

To sum up, then, designing a right-oriented deontologically grounded healthcare system involves several aspects: *First*, it means that the notion of act-effectiveness coefficient rather than health status outcome would carry the burden of establishing in calculative terms whether the policy structure of a healthcare system was in keeping with a deontological perspective. In other words, it means that a deontologically oriented healthcare system—which is to say, a healthcare system whose focus was equity rather than the greatest amount of common good—would contain the provision that if the act-effectiveness coefficient of a specific healthcare modality was not up to the standard that was achieved in other cases, then appropriate research would be initiated to try to remedy the situation. Only such a policy structure would satisfy the equity criterion and the Principle of Equality. However, as was said, such a policy structure would be a dynamic affair and would require a recursive assessment and adjustment as demographic, epidemiological and technological change.

Second, it means even though the healthcare modalities that would be included in the healthcare system might have the same act-effectiveness coefficients, that would not necessarily mean that the health status of their recipients would be the same, or even that their health status would be ameliorated, maintained or that it did not decline to the same degree. To argue thus would be to revert to the claim that an equitable distribution of healthcare modalities required a sameness of health status which, as has already been indicated, is not the case.

Third, it does not necessarily mean that if the act-effectiveness coefficient of a given healthcare modality was not the same as that of other healthcare modalities, then policy decision-makers could simply shift the funds from one modality to the other if doing so would likely lower the act-effectiveness coefficients of the other modalities and therefore would violate the Principle of Nonmaleficence.

[42] Pomare C, Churruca K, Ellis LA, Long JC, Braithwaite J. A revised model of uncertainty in complex healthcare settings: A scoping review. *J Eval Clin Pract*. 2019 Apr;25(2):176–182.

[43] Hatch S. Uncertainty in medicine. *BMJ*. 2017 May 11;357:j2180.

[44] Buetow S. The virtue of uncertainty in health care. *J Eval Clin Pract*. 2011 Oct;17(5):873–6

[45] Owens, D.K., Sox, H.C. (2001). Medical Decision-Making: Probabilistic Medical Reasoning. In: Shortliffe, E.H., Perreault, L.E. (eds) *Medical Informatics*. Springer, New York, NY.

Consequently, in order to stay within the overall deontological framework, another way would have to be found to satisfy the equitable treatment criterion of.

Clearly, there would be several ways of dealing with the situation without violating any deontological Principles. One such way would be to increase the funding that society accords to its public healthcare system. In other words, it would be to go back to the original macro-allocation process that deals with the society's resources and readjust the amount of resources that are accorded to healthcare in comparison to the other social services that the society is ethically obliged to provide.

However, while that is possible, it would run into the problem of having to decide the relative priority ranking of the various social services. This, of course, is possible, and it would take into account an aspect of health profiles that so far has been ignored: namely, the fact the act-effectiveness coefficients of many healthcare modalities is not simply a result of the nature of the relevant health conditions themselves or of the treatment modalities but is functionally related to the social embedding of the individuals who suffer from those conditions.[46] Moreover, it shows that ultimately designing an equitable healthcare system cannot be done independently of designing an overall system of social policies, for the issue of what amount of funds to allocate to healthcare itself is not part of the health policy-makers' mandate. If this is correct, then the data requirements for any AIs that the policy-makers would use in their deliberations would be expanded accordingly, and would once again have to straddle different domains.

Research Once More

However, that would still not be the end of the matter: Different healthcare modalities are differently effective, not only with respect to their nature but also with respect to the individuals who are served by the same modality. Considerations like this also fall under the mandate of health policy-makers.

There are many reasons for this. One is the life-style of the relevant individuals, another is their genetic endowment, and third is that many of the modalities have been developed over centuries of healthcare practice, some of them going back hundreds of years, where the most effective of these are not simply the result of tradition but are the result of dedicated research and investigation. It is the latter, then—i.e., it is the possibility of instituting a research programme and of correlating it with the relevant act-effectiveness coefficients—that provides the policy-makers with the way of dealing with this type of situation.

An initial task that would be part of the policy-makers' mandate, therefore, would be to divide the overall healthcare resources that are available to them into two rubrics: one that is intended to fund the healthcare modalities that are currently

[46] WHO, "Social Determinants of Health", https://www.who.int/health-topics/social-determinants-of-health#tab=tab_1

obtainable and to some degree effective, and another that is devoted to funding research for the health conditions for which no effective treatment modalities currently exist.

The reason that healthcare research would have to be contained as separate rubric is that while improving the effectiveness of a given healthcare modality may be the result of research that was conducted in a non-healthcare area—research in the physics of electromagnetic radiation such as x-rays would here be a good example—the results of such research do not necessarily or even directly affect the relevant healthcare modalities. Instead, they generally eventuate in the development or modification of a particular healthcare modality only as a result of scientifically valid research projects that focus on the healthcare condition itself. Therefore it is only after this allocation decision has been made that the policy-makers could proceed with the funding of the distinct modalities that met the equity criterion as it was outlined in the preceding discussion.

However, it also means that the ethical appropriateness of engaging in the relevant research projects would itself have to be measured in some way. Here the notion of act-effectiveness coefficients could again serve as a useful calculative tool. Thus, in order for a research protocol to be handled in an appropriate manner it would be necessary that the relevant piece of research would not be abandoned unless the act-effectiveness of the modality that was being investigated reached the level of the act-effectiveness-coefficient of the modalities in the existing healthcare system. This, in turn, means that some research would constitute an ongoing process, whereas other research could be terminated and its results be applied.

Therefore, integrating healthcare research into the policy framework in the above manner would satisfy the principles that define a deontologically oriented approach to healthcare while at the same time allowing for the fact that there may be differences in the effectiveness of the healthcare modalities that are actually being provided, and that the health status of the members of society may actually differ.

Implications for the Use of AIs

Of course the ultimate question so far as this study is concerned is whether—and if so, how—these considerations could be reflected in the design of an AI that might be used by the policy-makers.

Without going into the details of the actual algorithms of such an AI—a task that is more appropriately left to a purely technical investigation—the principle condition that would be relevant in this regard is that the various aspects that were outlined above could be translated into algorithms that mirror the logic of the preceding considerations. And without repeating the details of that logic, it seems valid to suggest that in principle this would present no difficulties because the reasoning that underlay these considerations was essentially logical in nature, and that appropriate algorithms could be designed as long as this was the case.

That aside, however, one of the fundamental requirements for the design and use of an AI would be that health the policy-makers, irrespective of their ethical orientation and irrespective of whether they viewed healthcare as a right or as a commodity, put in place a standardised health assessment form that would be completed for every member of society and kept up-to-date, so that thereby every identifiable aspect of the health condition of every individual would become a datum that could be used by an AI in the calculations that it undertakes. Given that AIs and their subroutines are electronic programmes, this means that each member of society would have to have an electronic health record that identified their health conditions.

Of course, as was already indicated in the preceding chapters, this means that, quite aside from ensuring that such records could also be used in hands-on as well as in institutionalised healthcare[47] which would have formatting implications, the establishment of such electronic patient records would have privacy and security implications.[48] Therefore the collection of the relevant data as well as their storage, communication and manipulation would have to be protected. However, given that only the data themselves would be involved in the relevant AI calculations at the policy level, it may seem reasonable to suggest that at least as far as the policy level was concerned this condition would be met if neither the records themselves nor the data contained in them were in identified form but were de-identified or anonymized.

Unfortunately, however, if current studies are any indication, then in principle it is possible to re-identify anonymous data on the basis of three distinct data points.[49] If that is correct, then the way to deal with this possibility would be to require that, rather than being de-identified or anonymized, the data be encrypted. While in principle even this would not ensure that they could not be re-identified—quantum computers could possibly crack any encryption[50]—the protection that would be afforded by something like public-key encryption algorithms would nevertheless be evidence that the best effort had been made to protect the identity of the individuals in question. Ideally, also, the data could be stored in a central or a secure data-base because that would not necessitate the establishment of a secure communication network in order to allow the relevant AIs to access the data that they would need in their calculations. This, in turn, would require that appropriate technical standard be established to meet this condition, and that any AIs that might be used in this connection would have to meet these standards.

---

[47] Health Information Technology: The Need to Know vs. Data Security. *Consult Pharm.* 2016 Sep;31(9):468–9.

[48] Cf. Chap. 3, pp. 48 et passim; Chap. 4, pp. 71 *et passim.*

[49] Lubarsky B, Re-Identification of "Anonymized" Data. 1 *Georgetown Law Technology Review.* 202 (2017).

[50] Cryptographers Take On Quantum Computers". *IEEE Spectrum.* 2009-01-01.

The Issue of Conflicting Requirements

Ethical vs. Legal Requirements

The discussion so far has been based on the assumption that healthcare policy-makers considered healthcare to be either a right or a commodity, and that they developed their respective policies on that basis.

However, that may be too simplistic an assumption. The fact is that, for a variety of reasons, the views that societies hold on the nature of healthcare itself tends to involve a combination of right- and commodity-oriented perspectives, and this translates into distinct and sometimes even incommensurable requirements with which the policy-makers have to deal. This, in turn, means that the policies that they have to establish may have to involve a variety of distinct and mutually inconsistent strands.

If this is the case, then it means that the AIs that the policy-makers could actually use in their deliberations could not have a unified logical structure, because a unified structure would be logically contradictory. The policy-makers would then have to use differently structured AIs, depending not only on the type of issue that they were considering but also on the perspective from which they had to approach the matter. Therefore, the AIs that the policy-makers could then use to assist them in drafting the relevant policies could not be of a single sort but would have to have distinct designs, each reflecting a distinct type of perspective. As to the decision about which sort of AI to use, ultimately that would lie in the hands of the policy-makers themselves. It would not be a matter of pure AI decision-making.

Funding Considerations

Which brings into focus a different issue: Namely, whether adopting a specific perspective on the nature of healthcare would rule out structuring the relevant policy framework by using a method of funding healthcare that was essentially at home in a different perspective. More specifically, it seems that commodity-oriented funding considerations are ruled out when healthcare is construed as a right. If this is correct, it would mean that the AIs that might be used by policy-makers when designing a right-oriented healthcare framework could not involve commodity-oriented calculations when assisting the policy-makers in their deliberations.

However, to argue thus would be to conflate considerations that deal with how healthcare itself is funded with the conditions under which healthcare may be accessed by individual healthcare consumers. To expand on this—and as was mentioned on several occasions—the implementation of a societal healthcare framework depends on the resources that the society has available. Consequently, the relevant policies that the policy-makers can realistically put in place have to move within that framework of limitation. This, in turn, means that the policy-makers have to consider what sorts of providers would be best suited to bring about the realisation of the healthcare policies within that framework.

These providers, however, fall into two distinct groups: They may be either commercial or non-commercial in nature. Furthermore, it may happen that under certain circumstances having the implicated healthcare services provided by a commercial provider is cheaper—i.e., is financially less resource-intensive—than providing it by means of a non-commercial public healthcare agency.

Consequently, if the policy-makers went about their task conscientiously, then they could not rule out the use of commercial providers. This would not mean that then they would have abandoned the perspective that healthcare was a right, nor would it mean that then their healthcare framework and its implementation would be contradictory. It would simply be a reflection of a fact that was pointed out before: namely, that the nature of a tool does not necessarily have the same logic as the reason why the tool is used. To illustrate this by an example that is not related to healthcare, the use of an explosive device such as a bomb to change the flow of a river so that it will provide water to an otherwise drought-starved population does not mean that one has thereby abandoned a peace-oriented orientation for a war-oriented orientation. Therefore, as long as the cost of accessing the healthcare modalities that such commercial sources provided did not fall to the users of their services and society itself paid for the actual services, the healthcare policy-makers would not need to eschew using commercial agencies to provide the services in question.

However, the AIs that the policy-makers might then use in reaching such a funding decision would have to contain subroutines that would allow them to calculate which method of funding the relevant interventions would be most economical and least resource-intensive. And that, once more, would expand the data requirements that the AIs would need in order to evaluate the relevant situations.

Conclusion

The purpose of the present chapter has been to outline how the overall structure of a healthcare system is functionally related to framework that is put in place by duly empowered policy-makers, and that this depends on whether healthcare is construed as a right or as a commodity. It was also outlined why and how the distinct ethical perspectives that the policy-makers adopt would result in different policies, as well as how some of these distinct parameters affect the design and functioning of any AIs that the policy-makers might use when carrying out their mandate. At the same time, however, it was suggested that to approach the policy-makers' task in this way may be more theoretical than realistic because, for various reasons, the mandate of the policy-makers may enjoin them to combine commodity- and right-oriented considerations in the framework that they put in place.

Moreover, the underlying assumption of the whole discussion—and one that was argued for in Chap. 1—was that when AIs are used in the design and the delivery of healthcare, they are used as tools: In other words, that they are what in Chap. 1 were called AI-2 s. In light of current developments in the field of artificial intelligence however, the functional as well as the material nature of AIs appears to be undergoing a series of fundamental changes that would essentially put them into the category of persons or agents, rather than into the category of tools. In other words, that they would then be what were called AI-1 s. This, of course, means that the considerations that were adduced in this and in the preceding chapters would either no longer apply to them at all or that they would not apply to them in the ways that they were suggested. It therefore seems appropriate to devote a separate discussion to this possibility: i.e., to the ethical issues that would arise if and when the AIs in healthcare function as agents or persons. This, then, is the subject of the final chapter: The ethical implications of AIs being healthcare decision-makers in their own right.

Chapter 6
The Future: Artificial General Intelligence in Healthcare

Abstract The present chapter deals with ethical considerations that arise if and when AIs change to AGIs. It suggests that if and when this occurs—and it sketches some conditions under which this may be expected—they would then have to be treated as persons, and the ethical considerations that were outlined in the preceding chapters would no longer apply to them derivatively but directly. Moreover, some new ethical issues would also arise: Issues that, by and large, have not even been considered.

Introduction

The preceding chapters have sketched some of the ethical issues that arise when AIs are used in healthcare at the hands-on, institutional and social policy level. These considerations, however, were based an assumption that was made in the introductory chapters: namely, that although the AIs that currently exist have great complexity, they are tools and not independent agents in their own right, and that therefore the ethical considerations that were raised in the subsequent chapters apply to them only derivatively. In other words, the analysis was based on the assumption that AIs as they currently exist are not strong or full AIs—what in Chap. 1 were called AI-1s or artificial general intelligences (AGIs)—for, they lack what is central to the direct ascription of ethical accountability: namely, personhood, which requires the capacity for awareness and volition.[1]

This does not mean that current AIs may not contain subroutines that allow them to select among different options. However, how they execute that selection is determined by their programming, and this programming controls all of their functioning. Their selecting, therefore, is not choosing in the volitional of sense that term, which involves awareness, but is merely an automatic and predetermined way of selecting among alternatives. Therefore, it is more accurately described as a causal

[1] Cf. Chap. 1, pp. 12 ff.

E.-H. W. Kluge, *The Ethics of Artificial Intelligence in Healthcare: From Hands-on Practice to Policy-making*, The International Library of Ethics, Law and Technology 25,
https://doi.org/10.1007/978-3-031-72837-2_6

sorting-out process on the basis of their input and of the criteria and procedures that have been built into their subroutines. An although that selection process may occur on the basis of a random process—for instance, if their design determines that it should occur on the basis of a random number generator—that selection, even when its outcome is essentially unpredictable, is still be determined because it occurs on the basis of subroutines that stipulate that it should occur in a particular way.

Therefore, AIs as they currently exist ultimately function like so-called Turing machines:[2] That is to say, when they are in a particular state, they will proceed to the next state on the basis of their current state, their input and the transition rules that have been built into them. This holds true for all current AIs, whether they are programmed into a single computer or into an array of distinct electronic devices.

Finally—and to follow up on what was already indicated in Chap. 2[3]—if the capacity to make decisions independently of being constrained is called autonomy or free will, and if the possession of a free will is necessary for the direct ascription of ethical predicates[4], and if only human beings currently have that capacity, then this would justify the claim that ethical evaluations apply directly only to human beings, and that they apply to AIs only in a derivative sense.

Human Beings, AIs, and Free Will

This, of course, immediately entrains the question why the claim that human beings have volition or free will should be accepted. Arguably, this claim is based on a prejudicial preference for biological entities in general and for human beings in particular. Consequently, if and when artificial intelligences become more complicated or are constructed of biological parts—bioprinting here comes to mind—then they would also be autonomous agents and be subject to direct ethical evaluation.

The Gödel Argument

However, a possible way of countering this argument and of defending the claim that human beings and AIs are fundamentally distinct was already indicated in Chap. 2: namely, by using what might be called the Gödel Argument.

As was then mentioned, Kurt Gödel had shown that there are statements in mathematics that are known to be true but that can neither be proved nor disproved, and

[2] Turing AM. On computable numbers, with an application to the Entscheidungsproblem *J. of Math*, 1936–140.

[3] Chapter 2, pp. 26 ff.

[4] Weisman D. (2018). "Autonomy and Free Will," *Metaphilosophy* 49:5 609–645.

that it is impossible to prove that mathematics itself is consistent.[5] As was also indicated, all subsequent attempts to show that Gödel's analysis is mistaken have failed. This point was then combined with Alan Turing's thesis that a purely material entity like a machine cannot go through an infinite iterative process in a finite amount of time. It was then argued that since Gödel's proof deals with the transfinite relationships between numbers, which are non-denumerably infinite, a purely material entity could not establish such a proof in a finite amount of time. This, so the argument concluded, showed that the human mind could not be a purely material or physical entity because Gödel established his proof in a finite amount of time. Consequently human beings are fundamentally distinct from AIs, which are purely material or physical entities. More specifically, human minds cannot be identical with human brains, because these are material in nature. Therefore human minds must be entities that belong to a different dimension that is not bound by the deterministic laws of material reality. That is why human beings have the capacity for volition and of exercising free will.

A possible objection to this line of reasoning, however—and one that suggests that human brains are simply biological AIs—is that a proper analysis of how the human mind works should take into account the interaction of the neurons that make up the human brain. After all, Gödel used his brain when he engaged in his reasoning. However, if what has been suggested by Hameroff and Penrose in this regard is correct, then neurons interact at the quantum level.[6,7] Therefore, if this correct, then the Gödel Argument does not really show that purely material beings cannot give rise to Gödel's proof. All it shows is that the argument that human minds are not purely material in nature has not proceeded far enough: That it should take into account the nature of neurons and of their interactions. Those interactions, however, because they occur at the quantum level, do not proceed in the manner of Turing machines, and are not subject to the deterministic laws that govern material objects.[8] Consequently, a proper analysis at the quantum level would explain how free will is possible for purely material entities. After all, having free will is nothing other than not being subject to such deterministic laws.[9] It is therefore unnecessary to hypothesize that human minds are not purely material entities.

Consequently, so the counterargument would continue, if AIs functioned at the quantum level—for instance, if they were programmed into in quantum computers[10]

[5] Stanford Encyclopedia of Philosophy, "Gödel's Incompleteness Theorems", accessed at https://plato.stanford.edu/entries/goedel-incompleteness/#Int.

[6] Hameroff S and Penrose R, Conscious Events as Orchestrated Space-Time Selections, N*euro Quantology* 2003; 1: 10–35.

[7] Hameroff S. (20012). How quantum brain biology can rescue conscious free will, *Front Integr Neurosci.* Oct 12; 6:93.

[8] Hidary JD. (2021. 2nd ed.) *Quantum Computing: An Applied Approach*, Springer

[9] Stanford Encyclopedia of Philosophy, Incompatibilist (Nondeterministic) Theories of Free Will; https://plato.stanford.edu/entries/incompatibilism-theories/ .

[10] Nielsen MA and Chuang IL. (2010). *Quantum Computation and Quantum Information*, Chapter 7; Cambridge University Press.

or consisted of bioprinted neurons—then they would also not function like Turing machines but function like human beings. Therefore, if the sort of analysis that is offered by Hameroff and Penrose is correct—and the debate on the matter continues[11]—it is arguable that as soon as artificial intelligences are programmed into quantum computers or consisted of bioprinted neurons then they would also have free will and volition. Consequently, such AIs would be also subject to direct ethical assessment or predication. And it would also follow that if such AIs operated in the field of healthcare, the various ethical considerations that were adduced in the preceding chapters with respect to human beings would apply to them as well.

AIs in Quantum Computers

To see whether this last argument is valid, it may be appropriate to take a closer look at what is involved in being a quantum computer.

Without going into the technical details, quantum computers are constituted of parts that do not function at what might be called the atomic or molecular level but at the sub-atomic level. As quantum theory explains, entities at that level are not subject to the Law of the Excluded Middle and the Law of Non-contradiction. They may be superposed, which is another way of saying that they may be in opposite states simultaneously. Consequently, they can be in both an "on" and "off" state at the same time;[12] and to mark this difference, computer chips that function in this manner tend to be called "qubits" rather than simply "bits". This—so the argument continues—means that if AIs that were programmed into quantum computers, they would then have free will. Consequently, such artificial intelligences would in fact be agents in the full-blooded sense of that term; and if they were engaged in the planning or delivery of healthcare, then they would not be mere tools but persons in their own right, their inorganic constitution notwithstanding. For that reason, also, such AIs would be directly subject to ethical considerations rather than in merely a derivative sense

[11] Boolos, George; et al. (1990). "An Open Peer Commentary on The Emperor's New Mind". *Behavioral and Brain Sciences*. 13 (4): 655; Davis, Martin (September 1993). "How subtle is Gödel's theorem? More on Roger Penrose". *Behavioral and Brain Sciences*. 16 (3): 611–612; Landau J, "Penrose's Philosophical Error" in L.J. Landau and J.G. Taylor eds. *Concepts for Neural Networks*, (Springer 1997); Tegmark, Max (2000). "Importance of quantum decoherence in brain processes". *Physical Review E*. 61 (4): 4194–4206; Hepp, K. (September 2012). "Coherence and decoherence in the brain". *Journal of Mathematical Physics*. 53 (9): 095222.

[12] Nielsen, Michael A.; Chuang, Isaac L. (2010). *Quantum Computation and Quantum Information*. Cambridge University Press.

Biologically-Based AIs

An analogous argument could be based on the hypothesis that the AIs in question are biologically constituted.

It begins by accepting the claim that despite having material brains that are composed of neurons, human beings are genuine free agents. However, as Hameroff and Penrose have suggested, the functioning of these neurons is explained by quantum analysis. Therefore their functioning may be at variance with the deterministic laws of nature. This is what allegedly accounts for the fact that human brains do not function like Turing machines and that human beings have free will. Therefore—so the argument continues—if AIs were to be constructed of biological components like those that make up human brains, then they would also have free will and they would also be voluntary agents. In other words, then they would then be persons or agents in their own right, and ethical considerations would apply to them directly rather than in a derivative fashion. Consequently, if they were engaged in healthcare, the considerations that were adduced in the preceding chapters would also apply to them.

As to how this might be achieved, arguably the answer lies in the technique of bioprinting: More specifically, in the technique of using living cells and biomaterials to print an object layer by layer to produce cells and organ-like structures that are living and that can multiply.[13]

The technique itself has been around for some time, and it has advanced to the stage where it can produce not only renal and cardiac cells but also neurons.[14] Not only that: once produced, these neurons can in principle not only be assembled into functioning neural networks but can also modify themselves both at the cellular and the network level, much as is the case with the brains of developing human fetuses. To be sure, the technology has not as yet reached this stage. However, given the speed and the direction in which the technology is developing—so-called neurobiochips are already being constructed[15]—it may only be a question of time until it is the case and the development of biologically-based AIs becomes a reality. Consequently, bioprinted artificial intelligences could then emulate what human beings can do, and would therefore have volition and free will. In other words, they would no longer be what in Chap. 1 were called AI-2s but AI-1s. And, as was

---

[13] Dey, M., Ozbolat, I.T. 3D bioprinting of cells, tissues and organs. *Sci Rep* 10, 14023 (2020); Massetti, M., Zhang, S., Harikesh, P.C. et al. Fully 3D-printed organic electrochemical transistors. *npj Flex Electron* 7, 11 (2023); Mobaraki M, Ghaffari M, Yazdanpanah A et al (2020) Bioinks and bioprinting: a focused review. *Bioprinting* 18:e00080.

[14] Knowlton S, Anand S, Shah T, Tasoglu S. Bioprinting for Neural Tissue Engineering. *Trends Neurosci*. 2018 Jan; 41(1):31–46; Antill-O'Brien N, Bourke J, O'Connell CD. Layer-By-Layer: The Case for 3D Bioprinting Neurons to Create Patient-Specific Epilepsy Models. *Materials (Basel)*. 2019 Oct 1; 12(19):3218; Liu X, Hao M, Chen Z, Zhang T, Huang J, Dai J, Zhang Z. 3D bioprinted neural tissue constructs for spinal cord injury repair. *Biomaterials*. 2021 May; 272:120771.

[15] NeuChip.eu, https://neuchip.eu/

suggested above, if such AIs were engaged in the provision of healthcare, then they would not be mere tools, any more than the medical colleagues or residents that carried out the instructions of an attending physician. They would be agents in their own right. Consequently, as was said, ethical considerations would apply to them directly, rather than in a derivative fashion.

Some Philosophical Considerations

However, it is unclear that either the quantum-computer argument or the bioprinting argument really establishes the point at issue: Namely, that if artificial intelligences were constructed in either fashion then, unlike artificial intelligences as they currently exist, they would have free will and therefore would be persons or voluntary agents. For arguably—and this is the important overarching point—it is not at all clear that the issue of free will is even relevant when deciding whether an entity can count as a moral agent.

The Issue of Free Will

More specifically, the issue of free will itself is and has been a matter of debate ever since the beginning of philosophy and of the discipline of ethics, and its analysis has varied from philosophical school to philosophical school—indeed, almost from philosopher to philosopher.

Thus—and speaking for the moment historically—metaphysical monists like Parmenides[16] in ancient Greece and Spinoza[17] in the Seventeenth century denied that there was such a thing as free will at all. They maintained that reality is an integrated deterministic whole and that human beings are merely aspects or parts of this one overarching reality. The notion of freedom and of free will applied neither to the parts of this reality such as human beings nor to reality considered as a whole because this all-encompassing whole was structured according to the laws of logic. Consequently—so they argued—all relationships and all actions, being subject to the laws of logic, were determined. Therefore free will did not exist and was essentially a metaphor.

---

[16] G. S. Kirk, J. E. Raven, M. Schofield. *The Presocratic Philosophers: A Critical History with a Selection of Texts*. Cambridge University Press, 1983.

[17] Spinoza, Bendict de. *Ethics*; available at https://www.gutenberg.org/files/3800/3800-h/3800-h.htm.

Moreover, historical materialists like the ancients Stoics[18] also denied that human beings have free will. And, switching to more modern times, so do contemporary materialist philosophers like P.S. Churchland[19] or J.A. Fodor.[20]

In opposition this, metaphysical dualists like Plato[21] and Descartes[22] maintained that human beings do have free will because—so they argued—while the material universe is subject to deterministic causal laws, immaterial or spiritual reality is not. Moreover—so they claimed—human beings are not purely material entities. Metaphysically they are bipartite entities that are constituted of both an immaterial mind and a material body. Therefore—so they taught—while the material universe was strictly deterministic in its functioning, the immaterial or spiritual universe of which the human mind was part was not. On this basis they insisted that because human minds were immaterial, they were not subject to the determinism that characterises purely material beings and therefore they did have free will. That is why they could be held ethically responsible for their decisions and actions.

Still other philosophers have simply assumed that human beings have free will. For instance, while Immanuel Kant analysed the ethical implications of free will, he simply assumed that human beings have free will, and he based his whole ethical speculations on that assumption.[23] Likewise, more contemporary ethicists—whether of a consequentialist or non-consequentialist orientation, and whether dualists or materialists in their metaphysics—do likewise, as is obvious from the works of philosophers like G.E. Moore,[24] J. L. R.M Hare,[25] and Peter Singer,[26] to name but a few more well-known individuals. They do not prove it but simply assume it. Other philosophers, as for instance J.L. Mackie, also simply assume that human beings have free will, which is why human beings can freely choose the values on which they base their decision-making.[27]

Moreover, it has even been claimed by some philosophers that the notions of free will and of determinism are not logically exclusive but are mutually compatible. So-called compatibilists[28] advance this perspective, and they maintain that both free

[18] Stanford Encyclopedia of Philosophy, *Stoicism*, available at https://plato.stanford.edu/entries/stoicism/.

[19] Churchland, P.S., 1986, *Neurophilosophy: Toward a Unified Science of the Mind/Brain*: MIT Press.

[20] Fodor JA, *Contemporary Materialism* (Routledge, 1995)

[21] Cooper, John M. (ed.), 1997, *Plato: Complete Works*, Indianapolis: Hackett.

[22] Stanford Encyclopedia of Philosophy, *René Descartes;* available at https://plato.stanford.edu/entries/descartes/.

[23] *Groundwork for the metaphysics of morals*; available at https://www.earlymoderntexts.com/assets/pdfs/kant1785.pdf.

[24] Moore G.E. (1903), *Principia Ethica*, Cambridge University Press.

[25] Hare, R. M. (1989). *Essays in Ethical Theory*: Clarendon Press.

[26] Singer P. (2011) *Practical Ethics*), Cambridge University Press.

[27] Mackie, J. L. (John Leslie) (1978). *Ethics : inventing right and wrong*: Penguin Books.

[28] "Compatibilism". *Stanford Encyclopedia of Philosophy*. https://en.wikipedia.org/wiki/Compatibilism#:~:text=Compatibilism%20is%20the%20belief%20that,nothing%20to%20do%20with%20metaphysics.

will and determinism characterise human beings. By implication, therefore, these theorists could insist that artificial intelligences that functioned at a quantum level would also be agents in their own right and therefore would be directly subject to ethical considerations and not merely in a derivative sense. The only issue for them would be how one talked about AIs, and the types of uses to which one put them.

As well—and to round out the spectrum of philosophical positions—there is also the position of ethical emotivism. It maintains that ethical statements are nothing more than sophisticated expressions of an emotional nature, and that they say nothing about the nature of the individuals about whom they are made.[29] In other words, ethical emotivism argues that making an ethical judgement is essentially equivalent to saying something like "Boo!" or "Yea!" Ethical emotivism therefore entirely ignores the question whether the entities about whom such judgements are made have free will. And while it accepts that ethical statements may be more or less appropriate, depending on the situation, it argues that this is ultimately a matter of the social or cultural framework in which such statements are made: In other words, that it is ultimately a matter of convention. Therefore, if this position is adopted then, irrespective of how the AIs are constructed, whether they have free will is ultimately a matter of culture or convention and not a matter of fact. Consequently, if this position is correct then it is irrelevant how AIs are constituted. Whether so-called ethical predication applies to them is simply a matter of emotion and convention.

On the other hand—and to extend the discussion beyond the philosophical sphere—the thesis that human beings do have free will is shared by most religiously oriented thinkers, whether they be of the Judaic, Christian, Islamic or any other persuasion, and this even extends across the ethnic and cultural spectrum. For, irrespective of whether the religious frameworks teach that reward and punishment are administered by an overarching agency or are simply the result of the structure of the universe,[30] their teachings on reward and punishment would make little sense without the assumption of free will.

The legal sphere is another domain of human interaction in which the notion of free will plays an integral role, and this is the case irrespective of any specific differences in the relevant legal systems. Thus, the legal systems of all nations include provisions that distinguish between culpable and non-culpable acts, where the former are alleged to be the result of a free and voluntary decision on part of the actor whereas the latter are said to lack this element of volition. Likewise, they distinguish between consent and lack of consent and— irrespective of how the issue arises and irrespective of whether the consent is informed or otherwise—they assume that the whole process of consent itself lies within the free and voluntary

[29] Stevenson CL, *Facts and Values: Studies in Ethical Analysis*, New Haven: Yale University Press; Blackburn S; 1993, *Essays in Quasi-Realism*, New York: Oxford University Press; Gibbard A. 1990, Wise Choices, Apt Feelings: *A Theory of Normative Judgment*, Cambridge, MA: Harvard University Press; Ridge M. 2014, *Impassioned Belief*, Oxford: Oxford University Press.

[30] Das P. and Sahu V. Freedom of the Will and No-Self in Buddhism. *J. Indian Counc. Philos. Res.* 35, 121–138 (2018).

power of the individual in question. This would not be possible if it was not assumed, for whatever reason, that human beings had the capacity for volition or free will.

Discussion

Given the lack of philosophical agreement about which ethical approach is appropriate for assessing human actions and, more importantly, given the lack of agreement about what an ethical assessment itself amounts to, it becomes relevant to ask whether a philosophical discussion of the question whether artificial intelligences would ever have free will and count as agents rather than as mere tools is even relevant.

However, to abandon a philosophically oriented analysis is to ignore the fact that ultimately all ethical assessments are philosophical in nature. More importantly still, it is to ignore that philosophical speculations—in particular those that have an ethical content—have important practical implications. Thus, irrespective of which philosophical theory is ultimately considered appropriate for interpreting the thrust of ethical assessments, the core concept that underlies all of them is that ethical predicates apply to human beings because human beings have the capacity for volition or free will. In other words, the core concept remains that of a voluntary agent. That, however, is a philosophical concept; which, in turn, means that philosophical considerations have not dropped out of the picture after all.

Moreover, while the overarching framework of a philosophical discussion of ethics may be theoretical in nature and may be spelled out in theoretical terms, it still has practical implications, not the least of which are legal in nature.

Thus, if AGIs come to exist, have free will and are recognised to be persons then, like all other persons, they would also have a series of legal rights and duties. Consequently, what on the surface seems to be a series of purely philosophical and theoretical considerations is not simply a series of idle speculations. In fact, the following questions then arise: If—and when—such entities enter the sphere of healthcare, would the analysis that was presented in the preceding chapters with respect to the human users of AIs then devolve onto these entities? Also, would it pose new problems: for instance, with respect to reward and punishment? And if it did, would it be legitimate to deal with these problems without in some way involving these entities themselves? After all, ethical issues that deal with human beings are not dealt with by ignoring what human beings think and by not considering how the relevant human beings would be affected.

Overarching Considerations

The ultimate issue, then, is whether once there are artificial general intelligences or AGIs—and irrespective of whether they are quantum-mechanically or biologically grounded—they would have volition or free will, and should they be considered to be persons?

A Definitional Preamble

At this juncture, and before addressing the issue, it may be appropriate to go back a step and clearly define what is meant by the term "artificial generative intelligence" and to distinguish it from the term "artificial general intelligence"; for they are frequently used interchangeably but have fundamentally distinct meanings.

Thus, the term "artificial generative intelligence" refers to a computer-based expert system that is capable of generating text, images, videos or other data using generative techniques or algorithms in response to prompts. It functions solely of the basis of the routines and subroutines of which it is constituted, and these often are essentially statistical in nature. Examples would here be chatbots like ChatGPT, Gemini, Claude, Llama, Ernie, and Grok. Their programmes make them capable of so-called deep learning, where a series of multi-layered subroutines enable them to produce an output that simulates the product of human reasoning. They cannot, however, and do not fundamentally modify the structure of their computational frameworks, nor do they produce output independently of being prompted. Moreover, the types of prompts to which they respond are determined by and, so to speak, written into their programming.

The term "artificial general intelligence", however, refers to a computer-based operating system that has all the capabilities of an artificial generative intelligence but that is not limited in its functioning by the programmes that are built into its initial design.[31] Instead, it can modify the design of its routines and subroutines, as well as the values or action potentials that it assigns to its input and according to which its subroutines function. In other words, it is able to modify its processing programmes and deal with its input in an entirely new fashion. Therefore it is capable not only of making the output of its routines and subroutines ever more precise but also of performing tasks that were not in its original programme. As well, it does not need to be prompted to initiate an action: It can function independently. Arguably, therefore, an AGI would be conscious, self-aware and have volition. In short—material basis aside—it would essentially satisfy the criteria that are definitive of being a person.

[31] Heaven WD. (2023). Google DeepMind wants to define what counts as artificial general intelligence. *MIT Technology Review.* https://www.technologyreview.com/2023/11/16/1083498/google-deepmind-what-is-artificial-general-intelligence-agi/.

The Criterion of Personhood

However, putting the matter this way once more complicates the issue. For, as was already pointed out in Chap. 1, the concept of a person—i.e., the notion of personhood—is not a unitary concept but comes in two forms: namely, that of a natural person and that of a constructive person.[32] Natural persons, such as individual human beings, are entities whose material basis consists in biological bodies, whereas constructive persons such as corporations or collectivities are entities that consist of several individual natural persons who are associated with each other in certain ways.[33] Both types of persons are recognised in law and in ethics, and both types of persons have rights and obligations.[34] Therefore, once more, who or what counts as a person is not a purely theoretical matter but has serious practical implications.

At first glance, it would therefore seem that AGIs would fall into the category of constructive persons. This is because they would be constructed and composed of a number of components—where these may even be in different locations—and those components communicate with each other by electronic means and function in a coordinated manner. In that sense, therefore, they would seem to be analogues of corporations or collectivities.

However, closer consideration suggests that this would not really to be the case. Assuming for the moment that such AGIs satisfied the criterion of volition—which, after all, is the central point at issue—they would not have the fundamental characteristic of constructive persons as this was outlined above because they would not be constituted of parts or entities that are persons in their own right. And while the individual persons who make up collective persons like corporations and collectivities are also composed of parts—namely, of cells, organs and organ systems—the fact remains that these individual parts or components of constructive persons are actual persons in their own right.

However, this would not rule out the possibility that bioprinted AGIs would be persons. As was mentioned before, the technology is advancing in the direction of bioprinting neural networks that can be assembled—indeed, that can assemble themselves on the basis of the directives that are contained in their cellular makeup—into systems which, when supplied with the appropriate means, can do whatever human beings can do. As to the matter of requiring appropriate means to engage in such actions, in this they would be no different from human persons, who also cannot engage in a particular action unless they have the means to do so.

---

[32] Cf. Chap. 1, p. 16.

[33] Jovanovic, Miodrag, Constructing Legal Personality of Collective Entities – The Case of 'Peasants' (April 28, 2015). Available at SSRN: https://ssrn.com/abstract=2599926 or https://doi.org/10.2139/ssrn.2599926.

[34] Blair MM. (2013). Corporate Personhood and the Corporate Persona. *University of Illinois Law Review*. 785.

Therefore it would seem that once AGIs are constituted of biological components—as is currently being explored by NEU-CHIP[35]and similar projects—they would indeed count as persons. And this, in turn, would mean that not only would ethical assessment of their activities be appropriate, it would also mean that they would have legal standing and consequently would have rights as well as duties. Moreover—and with due alteration of detail—it would mean that all of the considerations that were raised in the preceding chapters relative to human beings in the conduct and planning of healthcare would apply to them as well.

The Issue of Volition

However—and leaving aside for the moment the issue of rights and duties that was mentioned above—it would still be premature to claim that this would settle the issue whether AGIs would then count as persons. For, it is not their constitution that defines a person in an ethically and legally relevant sense but their functioning as a voluntary agent. Thus, even when the individual neurons of a human being are replaced by chips—as currently being explored[36]— such an individual does not cease to be a person simply because their neurons have thus been replaced. If they were persons before and if they continue to function as they did before, then they continue to count as persons. It would be different if they then functioned as automata. Then neither an ethical nor a legal assessment of their actions would be appropriate[37] because they would then lack the capacity for volition or free will. However, as was said, that is not the case. This is clearly demonstrated by the fact that the individuals who undergo neuron replacement therapy still have the capacity for volition and do not function as automata.[38] It is this capacity, then, that identifies them as voluntary agents. And it is with this, that the issue once more returns to the question whether volition and free will would be—or could be—be found in AGIs.

In order to answer this question, therefore, it would be helpful if one could model the reply on the basis of how the capacity for volition and free will is established in the case of human beings. However, as the initial discussion has indicated, there is no agreement even here. In practice, of course, it is generally agreed that human beings do have free will and therefore—such conditions as automatism being

[35] NEI-CHIP.eu; https://neuchip.eu/eu

[36] Keene, Scott T.; Lubrano, Claudia; Kazemzadeh, Setareh; Melianas, Armantas; Tuchman, Yaakov; Polino, Giuseppina; Scognamiglio, Paola; Cinà, Lucio; Salleo, Alberto; van de Burgt, Yoeri; Santoro, Francesca (September 2020). "A biohybrid synapse with neurotransmitter-mediated plasticity". *Nature Materials*. 19 (9): 969–973; Oxley, TJ; Yoo, PE (2020). "Motor neuroprosthesis implanted with neurointerventional surgery improves capacity for activities of daily living tasks in severe paralysis: first in human experience" (PDF). *Journal of NeuroInterventional Surgery*. 13 (2): neurintsurg-2020-016862.

[37] Levy N, Bayne T. Doing without deliberation: automatism, automaticity, and moral accountability. *Int Rev Psychiatry*. 2004 Aug; 16(3):209–15.

[38] Grade, S., Götz, M. Neuronal replacement therapy: previous achievements and challenges ahead. npj Regen Med 2, 29 (2017)

excepted—they are responsible for what they do. This is even enshrined in the judgements of such institution as the International Court of Justice.[39]

Unfortunately, however—as was pointed out—even how the notion of free will itself is actually applied is relative to the assumptions that underlie the conceptual and cultural framework within which such judgements are made. And—to expand on what was said before—it is generally assumed in most cultures that even though young children are capable of exercising volition and have free will, they are not considered to be morally responsible for their actions because they lack sufficient knowledge, cognitive capacity and understanding to be able to exercise that will, and that is why they cannot legitimately be held to be morally responsible for what they do.[40]

However that, too, is merely an assumption and not an explanation. And while it ties the notion of free will to the notion of cognitive capacity and understanding, it is also generally agreed that these by themselves do not entail free will. This is amply illustrated by the fact that an *idiot savant*—who in many ways is similar to a biological AI in their actions—is not held to be morally responsible for what they do even though they may possess vast amounts of knowledge and can reason in a logically superb manner.[41] Therefore the possession of knowledge and the ability to reason do not seem to be sufficient to count as voluntary agent for that, as was said, can be accounted for in terms of data awareness and of data processing.

The Issue of Self-awareness

This, of course, raises the question whether awareness holds the key; and in particular whether it is self-awareness, which involves the functional interaction among neural systems in the human brain.[42]After all, a human child progresses from only incipient self-awareness while it is a foetus—at which stage it is only reactive to

[39] International Court of Justice, available at https://www.icj-cij.org/rules

[40] Bou-Habib, P. and Olsaretti, S., 2015, "Autonomy and children's well-being," in *The Nature of Children's Well-Being,* Dordrecht: Springer, pp. 15–33; Hannan S, 2018a, "Childhood and autonomy," in *The Routledge Handbook of the Philosophy of Childhood and Children*, Routledge, pp. 112–122; Levinson M, 1999, *The Demands of Liberal Education*, Oxford University Press; Mullin A, 2007, "Children, Autonomy and Care," *Journal of Social Philosophy*, 38(4): 536–553, and 2014, "Children, Paternalism and the Development of Autonomy," Ethical Theory and Moral Practice, 17(3): 413–426; Oshana, Marina, 2005, "Autonomy and Self-Identity," J. Christman and J. Anderson (eds.), *Autonomy and the Challenge to Liberalism*, Cambridge University Press; Ross L, 1998, *Children, Families and Health Care Decision-Making*, Oxford University Press.

[41] Treffert DA (May 2009). "The savant syndrome: an extraordinary condition. A synopsis: past, present, future". *Philosophical Transactions of the Royal Society of London. Series B, Biological Sciences*. 364 (1522): 1351–7.

[42] Lou HC, Changeux JP, Rosenstand A. Towards a cognitive neuroscience of self-awareness. *Neurosci Biobehav Rev.* 2017 Dec; 83:765–773.

stimuli—to awareness at birth.[43] Moreover, it appears that the development of self-awareness itself comes in stages, initially being only minimal but progressively increases as the child gets older and its brain develops.[44] Self-awareness is also tied to the development of the ability to adopt an ethical stance towards others, as is indicated by the fact that the ability to make ethical judgments progresses from by age 3 to age 5.[45] Arguable, therefore, if and when AGIs are constructed in a way that mirrors this functional complexity, then they too would—or should—count as persons.

AGIs as Persons: Some Ethical Issues

However, even if it was assumed that these issues were somehow settled and that AGIs did count as persons, that would not be the end of ethical issues that would arise in their regard. For, as was indicated before, persons not only are responsible for their actions, they also have rights.

Rights

In particular, the *Universal Declaration of Human Rights* stipulates that all persons have the right to life, liberty and security of their person.[46] It would therefore seem that this would entail that destroying; switching off or otherwise terminating such an AGI would be unethical.

Of course it might be argued that applying such concepts to AGIs would be too quick a move: It would be to ignore the fact that, as the title of the *Universal Declaration of Human* Rights itself as well as the formulation of its various Articles indicates, the rights that are mentioned belong to human beings.

However, to argue thus would be to simply decide that because AGIs would not be human beings, anything that was mentioned in the *Declaration* would not apply to them. Therefore, it would be to simply settle the issue whether AGIs that were engaged in the delivery or planning of healthcare would be subject to ethical

---

[43] Lagercrantz, H., Changeux, JP. The Emergence of Human Consciousness: From Fetal to Neonatal Life. *Pediatr Res* 65, 255–260 (2009)

[44] Oschwald J, Guye S, Liem F, Rast P, Willis S, Röcke C, Jäncke L, Martin M, Mérillat S. Brain structure and cognitive ability in healthy aging: a review on longitudinal correlated change. *Rev Neurosci.* 2019 Dec 18; 31(1):1–57; Lebel C, Deoni S. The development of brain white matter microstructure. *Neuroimage.* 2018 Nov 15; 182:207–218.

[45] Rochat P. Clinical pointers from developing self-awareness. *Dev Med Child Neurol.* 2021 Apr; 63(4):382–386.

[46] *Universal Declaration of Human Rights,* Article 3; available at https://www.un.org/en/about-us/universal-declaration-of-human-rights

evaluation by way of decision. Moreover, it would merely expand the problematic beyond the present context. For, if that decision was accepted, it would also raise the issue of animal and environmental rights—topics which themselves, rather than being settled, are matters of ongoing debate.

That aside, however, another issue that is entailed by the question whether AGIs would count as persons is the question whether they would then have a right of choice: i.e., whether the Principle of Autonomy would then apply to them, and whether they would therefore have the right to decide on their area of activity or practice. In other words, and among other things, it would raise the question whether AGIs that were active in the field of healthcare would have the right to abandon that type of activity at their discretion and, for example, go into to law or to work in construction. More particularly still, it would raise the question whether to design and fashion an AGI in such a way that it would inevitably function in a particular area would be ethical. And while the answer seems fairly clear as far as AI-2s are concerned—after all, they are tools, and ethical obligations are not owed to tools—it is not so clear what the answer would be in the case of AGIs if and when they should ever exist and count as persons.

More specifically—and leaving aside purely legal and social matters—all persons are assumed to have the right to decide what profession they will choose. Indeed, whether they will choose to pursue a profession or to engage in any other type of activity. Consequently, if AGIs counted as persons, then according to the Principle of Equality they would also have such a right. That, however, would mean that it would be unethical to design or structure an AGI in such a way that it would work in the area of healthcare without itself actually choosing to do so. That would be logically equivalent to genetically engineering or programming a human person in such a way that they would only engage in a specific occupation or type of activity. That, however, would be a violation of the Principle of Autonomy.

Furthermore, it is generally assumed that the individual or individuals who play the role of parents to a human child have an obligation to support that child, and that the society in which that child exists or comes into being has a corresponding obligation to put in place measures that will ensure that this happens so that the new person's needs will be taken care of.[47] Consequently, it is arguable that if AGIs count as persons, then the constructors of AGIs as well as the society in which that construction takes place would have an analogous set of obligations as the AGIs come into being.

Another ethical issue that would arise and that would have to be solved if AIs in the field of healthcare were AGIs would be the matter of recompense—and, once more, it is an issue that would arise for AGIs in general. Thus, it is generally agreed that all individuals have the right to be remunerated for the work that they do, and that not to remunerate them for their work and yet to expect them to engage in the relevant activity is a violation of the Principles of Equality and Autonomy. Thus, the

[47] United Nations, *Convention on the Rights of the Child*, https://www.ohchr.org/en/instruments-mechanisms/instruments/convention-rights-child

Universal Declaration of Human Rights, in Article 23, stipulates that "Everyone, without any discrimination, has the right to equal pay for equal work."[48] Consequently the question arises what form this recompense would take for AGIs, and what sort of provisions would have to be made in order to satisfy this right. And specific to AGIs in the field of healthcare, how would they be remunerated for whatever they did?

Of course, as was indicated before, it is arguable that the clauses of the *Declaration* mention only human beings, and that therefore they would not apply to AGIs. However, as was also said before, to claim that therefore they would not apply to AGIs would essentially be to decide that only human beings count as persons. That, however, would only settle the matter by decree and not by reasoned argumentation. Moreover, it would be to ignore the possibility that AGIs may find their origin in bioprinting with humanly based materials.[49] That, however, would make AGIs biologically human except for their method of origin. As to the significance of the method of origin, given the evolving techniques of *in vitro* fertilization and the development of artificial wombs[50], any such argument would also require independent justification.

However, dealing with these and similar issue transcends the scope of the present investigation and therefore will be left to another occasion. At this juncture, suffice it to say that any solution that was grounded in the concept of personhood would have to be based on an analysis that did not automatically and preanalytically distinguish between human beings and AGIs. Consequently, if ethical considerations are relevant with respect to genetic manipulation, education and support of human beings then, with due alteration of detail, these considerations would also apply to AGIs. And this would be the case not only with respect to their actual existence but also with respect to the various stages of their construction or coming-into-being, because that would be functionally equivalent to the gestation and development of human beings. Therefore, analogues of the issues of abortion and deliberate death as they pertain to human beings would also arise with respect to AGIs.

Duties

Another question that would have to be addressed if AGIs in the field of healthcare were persons is whether they would have responsibilities and, if so, what these would be.

That is to say, the AIs that are currently used in the field of healthcare are merely tools or expert systems. Consequently, any responsibilities that obtained in this regard would belong to the professionals who use them. This is because

[48] https://www.ohchr.org/sites/default/files/UDHR/Documents/UDHR_Translations/eng.pdf.

[49] Dey M, Ozbolat IT. 3D bioprinting of cells, tissues and organs. *Sci Rep*. 2020 Aug 18; 10(1):14023.

[50] Gibson RB, Campo-Engelstein L. Artificial Womb Technology and the Restructuring of Gestational Boundaries. Am J Bioeth. 2023 May; 23(5):106–108.

responsibilities in an ethical sense only belong to the tool-users, not to the tools that they use. Therefore, while it may reasonably asked whether an AI functioned well or was efficient in its operation, etc., it would be inappropriate to ask whether it functioned responsibly or fulfilled its duties in an ethically appropriate manner in any but a metaphorical sense.

However if, as was suggested, AGI would have the status of persons, then they would have an array of duties just like any other kind of person who worked in their respective occupation. Consequently, the duties that belong to human healthcare professionals would then belong to the AGIs as well, and these duties would belong to them even if the interaction between them and the patients was incepted or took place without the intervention of human intermediaries.

Consent

To put this into context, one of the more important obligations or responsibilities that belong to human healthcare professionals is that of obtaining informed consent for any intervention in which they propose to engage with their patients. Thus, as the International *Code of Ethics* of the World Medical Association states in clause 13,[51]

> In providing medical care, the physician must respect the dignity, autonomy, and rights of the patient. The physician must respect the patient's right to freely accept or refuse care in keeping with the patient's values and preferences.

How precisely this clause has to be met varies from jurisdiction to jurisdiction. However, it is generally agreed that unless it is impossible to obtain such consent and time is of the essence, a duly empowered substitute decision maker is not reasonably available and there is no reasonably available advance directive—and, of course, unless not performing that intervention constitutes a serious risk to the health of other persons[52]— informed consent is ethically required for any healthcare intervention

It might be argued that the issue of consent should be handled differently in the case of AGIs when they are interacting with human patients because AGIs are constructed entities: In other words, because they essentially are machines. Therefore—so it might be argued—this constitutes a fundamental difference in the framework of the relevant interaction. After all, one of the normal aspects of a healthcare professional-patient interaction as it currently exists is that the healthcare professional who is involved is a human being. While that professional may use an expert system or an AI in their decision making and in the process of their intervention, the framework of the interaction is human-to-human in nature. However, that would not

[51] WMA, *International Code of Medical Ethics*, accessed at https://www.wma.net/policies-post/wma-international-code-of-medical-ethics/

[52] Beauchamp TL and Childress JF, (2001), *Principles of Biomedical Ethics* (OUP), pp, 77–89 *et pass.*

be the case if it was an AGI that interacted with the patient. The framework of the interaction would then be human-to-AGI. And—leaving aside for the moment the possibility that the AGI might be bioprinted—the framework of the interaction would then be fundamentally different because the AGI would be a constructed entity or a machine. Therefore it seems that a human healthcare professional would have to become involved in order to obtain the relevant consent, the fact that the AGI might independently carry on from there notwithstanding.

However, it is arguable that to reason thus would be to ignore the ethics and the logic consent. That is to say, consent is ultimately a person-to-person affair. Certainly, in the context of healthcare, the mere signing of a form is ethically insufficient.[53] In other words, the individual who gives consent and the individual who receives the consent must be persons. Therefore, the central issue in consent is not is not how the individuals who are involved came into existence or how they are constituted, but whether they are persons. Therefore, if they are persons, then—as the Principle of Equality entails—they are subject to the same ethical considerations. Consequently, contrary to the initial inference, if the AGI in question had the status of a person, then the framework of their interaction with patients would be a person-to-person framework, and consent should be requested by the AGI and should be given to it. Therefore, no human healthcare professional would have to become involved because the requirements of consent—that it be person-to-person—would be met.

To be sure, the interaction between the AGI and the patient might have a different emotional tenor than a human professional-patient interaction. That, however, would also not legitimate the claim that the ethics of the consent process should be handled differently, any more than would the fact that a patient might react differently towards a healthcare professional who was of a particular ethnicity, sex or gender would licence the conclusion that the informed consent process should have a different ethical structure. The details of the communication process that might be involved or the counselling that would be appropriate to deal with such a situation might well be different. The ethical framework of the consent process itself, however, would remain the same.

Qualification and Licensing

Another obligation—and one that has not previously been mentioned except in passing and in the context of human healthcare professionals—is that, both ethically and legally, a human healthcare professional is duty-bound to be properly qualified and licensed in the area in which they practice. Arguably, therefore, if AGIs were considered to be persons and if they functioned in the delivery of healthcare, this requirement would also apply to them: They would also be required to exhibit appropriate qualifications and be duly licensed before they could practice in their

[53] Finkelstein D, Smith MK, Faden R. Informed consent and medical ethics. Arch Ophthalmol. 1993 Mar; 111(3):324–6.

particular area of healthcare. The issue here is how, in their case, that obligation could be satisfied.

The usual way in which this requirement is satisfied by human agents is by their acquiring a certain amount and type of education and then passing a specialised series of exams and tests that establish their competence.

With due alteration of detail, these requirement could easily be satisfied by an AGI by making sure that the subroutines that the AGIs employed when engaged in the relevant activities would be at least as efficient and result in at least as accurate conclusions as the diagnostic, prognostic and interactive procedures that would be used by their human counterparts. This, in turn, could easily be achieved by designing the AGI in such a way that it was capable of deep learning,[54] by instilling the relevant data into their subroutines, and by designing their subroutines in such a way that they would satisfy the efficiency and accuracy requirements that have to be met by their human counterparts. Moreover, if the available evidence is any indication, this could easily be achieved. In fact, it seems that even AI-2s as they currently exist perform at least as accurately and effectively as the average human healthcare professional in the particular area for which they are designed. Therefore, it would seem that programming AGIs in a similar manner would also have the desired effect.

However, while efficiency and operational effectiveness would not present any technical problems in the design and construction of such AGIs, it is not clear that this would not present an ethical problem. For, as was pointed out before, if they were to be considered to be persons, this would violate their ethical status. More specifically, to design and construct them in this manner would then be to predetermine that they would function in a particular type of endeavour and in a particular manner. That, however—to focus on AGIs in the area of healthcare—would not give them any choice over whether they would function in the field of healthcare. However, as was previously pointed out, this would be unethical because it would violate the Principle of Autonomy, which would apply to them if they were persons. To draw an analogy, it would be like genetically engineering a human being in such a way that they had no choice but would inevitably become a nurse, a physician or other type of healthcare professional and would function in a particular way.

The only way to avoid this would be by not designing and constructing them as *healthcare* AGIs but simply as general AGIs whose subroutines did not include particular sets of data and were not specific to any given area of functioning. In other words, if AGIs were considered to be persons, their initial structure would have to be essentially undetermined with respect to any particular area of endeavour and without any inherent store of data. Their design would have to involve only subroutines that would allow them to acquire data and to acquire techniques of functioning. In other words, they would have to be able to choose the particular area in which to engage, and they would have to be able to acquire the data and techniques that would enable them to meet the relevant proficiency requirements of their

[54] LeCun, Yann; Bengio, Yoshua; Hinton, Geoffrey (2015). "Deep Learning". *Nature*. 521 (7553): 436–444.

chosen area of endeavour. Therefore, it would mean that the AGIs would have to designed and constructed in a fundamentally different manner from the way in which current planning envisions the design and construction of AGIs. For, that approach deals with them as though they were merely more sophisticated AIs or expert systems: In other words, it does not treat them as persons.

Liability

Moreover, if (and when) AGIs actually qualify as persons and function as healthcare professionals, another issue that arises would be that of liability.

That is to say, someone who has a particular responsibility is considered to have failed in carrying out their duty if they fail to act in an appropriate manner. They are then held to be liable for any damage, harm or misadventure that results from such a failure and they are obliged to provide suitable restitution to the affected person for any damage, harm or misadventure that resulted from the failure in question.

Moreover, this is not merely a legal matter but also extends into the area of ethics. Ethically, a professional is held to be responsible for what they do or fail to do in their interactions with their clients. Furthermore, this is held to be true irrespective of the nature of the profession. Therefore it holds for healthcare professionals as much as for lawyers. Consequently, if a healthcare professional fails in their responsibilities, they may be held liable by their patients for any damage, harm or misadventure that results from any such failure, and they may be required to provide the patients with appropriate recompense or restitution. Consequently, if an AGI functioned in the role of healthcare professional and failed in any aspect of the responsibilities that would normally be associated with a human healthcare professional who acts in such a capacity, this would also constitute a failure of duty on part of the AGI and therefore would also entrain liability.

In the case of human healthcare professionals any such matter is usually dealt with by means of disciplinary procedures, legal proceedings or by both, and it may result in consequences such as the professional being disqualified, having to undergo retraining or in a variety of other appropriate measures. It may even be the case that the affected patient (or patients) would be accorded financial damages. The question, therefore, arises how all of this would be (or could be) reflected in the case of healthcare AGIs.

Assuming for the moment that the healthcare AGI was licensed, the possibility of disciplinary procedures would present no technical difficulties: Analogous measures could consist in restricting the AGIs' area of practice, in requiring an improvement in their processing algorithms, etc. The AGI could even be given the option of withdrawing from or being engaged in healthcare altogether.

That aside, however, the serious problem that would here arise lies in the issue of restitution. More specifically, it lies in the issue of how and in what manner the AGIs would be able to satisfy the requirement of providing any affected patients with appropriate restitution if they did not have access to fungible goods or to their equivalent and restitution should be required. Therefore, if the AGIs did not receive a suitable recompense for their work, they would not be able to provide any restitution that might be required.

This, in turn, is part of a much larger issue: Article 17 of the *Universal Declaration of Human Rights* stipulates that "Everyone has the right to own property alone as well as in association with others."[55] Moreover, not to be recompensed for their labour either means that unless the individuals in question were engaged in volunteer work—an issue that itself would require further analysis and justification in the case of AGIs—to require them to perform work without appropriately recompensing them is essentially to turn them into slave labour. However, as Article 4 of the *Declaration* states, "No one shall be held in slavery or servitude; slavery and the slave trade shall be prohibited in all their forms."[56]

The relevance of these considerations with respect to the notion of AGI liability lies in the matter of recompense that was mentioned above. As was then pointed out, a patient may be entitled to receive recompense for any injury or misadventure that befalls them as a result of the professional's failure to act or failure to act in a professionally appropriate manner. However, that would be impossible if the professional—in this case, the AGI—did not possess any fungible assets. Consequently, if AGIs were considered to be persons and were to function in a healthcare capacity then, unless they chose to work for free—as do some human healthcare professionals who are members of such institutions as *Médecins Sans Frontières*—they would also have to be recompensed for their work. Only in this way would they be able to deal appropriately with any liability that they might incur.

Therefore, once more, unless it was simply decided that the rights that are outlined in the *Declaration* belonged only to human beings—and, as was mentioned before, that would require independent justification—the question arises in what sense healthcare AGIs could be recompensed for what they did, and what the nature of that recompense would be. Currently, this issue has not even been addressed in the various attempts to design and construct AGIs, whether that be in the area of healthcare or in any other domain. Moreover, it is unclear how—and indeed whether—appropriate provisions could be made for AGIs.

The Fiduciary Relationship

Moreover, there is also the issue whether and how a fiduciary relationship can come about in an AGI-human-patient interaction—and this goes to the very heart of the ethics of AGIs in healthcare

In general terms, such a fiduciary relationship does not come into about simply because one individual is qualified as healthcare professional and another requires a healthcare. Thus, the mere fact that one individual at a social gathering is qualified as a healthcare professional and another individual requires healthcare does not

[55] Op. cit. Article 17.

[56] Op. cit., Article 4.

entail that the former stands in a fiduciary relationship to the latter. Neither does the fact that a healthcare professional is in the same lavatory as someone who has a health issue establish a fiduciary relationship on part of the former towards the latter. It is the context of their interrelationship that determines whether a fiduciary relationship comes into being.

The context, in turn, is the overarching framework in which the two find themselves. If that framework is one of health care delivery, and if the professional is functioning *as* a healthcare professional, then the professional acquires a fiduciary obligation towards the relevant individual when they latter interacts with them. Likewise, if the professional interacts with another individual on a professional basis but not in an institutional or official setting—for instance, if they visit them in the latter's home but in a professional capacity—then a fiduciary relationship also arises. However, if the healthcare professional interacts with the individual on a purely social basis or is in the mere presence of such an individual—for instance, in the passenger cabin of an airplane—a fiduciary relationship would not arise. In such a situation, a fiduciary relationship would exist only insofar as there is a humanitarian duty to provide care, as is indicated by the World Medical Association's *Code of Medical Ethics* which states:[57]

> A PHYSICIAN SHALL give emergency care as a *humanitarian* duty unless he/she is assured that others are willing and able to give such care.

However, that would not mean that a fiduciary existed on its own. Even when the professional was following their humanitarian duty, a fiduciary relationship would come into being only when they actually interacted with the other person on a professional basis. The question, of course, is whether a humanitarian duty could be ascribed to an AGI and, more particularly still, whether a humanitarian duty could be ascribed to an AGI who was qualified as a healthcare professional and found itself in such a situation. Consequently, it is at least arguable that the domain of healthcare ethics would have to be completely restructured if and when healthcare AGIs actually come to exist.

Conclusion

The preceding discussion has attempted to sketch some of the ethical considerations that would arise if AGIs in healthcare were more than very complex expert systems and tools that are used by human healthcare professionals but were person or actual agents in their own right. It has also tried to identify and discuss some of the presuppositions that would have to be met in order for this to be the case with respect to

[57] WMA *Code of Medical* Ethics, "A PHYSICIAN SHALL give emergency care as a humanitarian duty unless he/she is assured that others are willing and able to give such care", emphasis added; https://www.wma.net/wp-content/uploads/2006/09/International-Code-of-Medical-Ethics-2006.pdf

their material constitution and functioning, as well as with respect to the ethical considerations that would then be relevant. Among other things, it was suggested that not to abide by the same ethical principles in their case as those that apply to human persons would be discriminatory. This was then spelled out in terms of rights and duties. It was also suggested that, with due alteration of detail, all of this would also have legal implications.

As was said, currently there are no AGIs. Therefore the issues that were discussed in the preceding paragraphs exist only as hypotheticals. However, if current technical developments are any indication, AGIs may very well come to exist—some technical developments seem to point in that direction—and then the issues will cease to be merely hypothetical and become real. Of course in practical terms, these ethical issues can always be handled by legal means or even by ignoring them, just as is the case with many issues that touch human beings. But to handle them in that way is to ignore their fundamental nature.

Finally—and as the preceding discussion has attempted to indicate—the ethical issues that would arise could not be handled by purely technical means without first establishing that these means would not only be effective but would be ethically defensible. For, as Kant already pointed out hundreds of years ago, the fact that something works does not in itself mean that it is ethical.

The manufacturer's authorised representative in the EU is Springer Nature Customer Service Centre GmbH, Europaplatz 3, 69115 Heidelberg, Germany. If you have any concerns regarding our products, please contact ProductSafety@springernature.com

Printed and bound by CPI Group (UK) Ltd, Croydon, CR0 4YY
01/10/2026
02225876-0013

SPRINGER NATURE

GPSR Compliance

The European Union's (EU) General Product Safety Regulation (GPSR) is a set of rules that requires consumer products to be safe and our obligations to ensure this.

If you have any concerns about our products, you can contact us on ProductSafety@springernature.com

In case Publisher is established outside the EU, the EU authorized representative is:

Springer Nature Customer Service Center GmbH
Europaplatz 3
69115 Heidelberg, Germany

The manufacturer's authorised representative in the EU is Springer Nature Customer Service Centre GmbH, Europaplatz 3, 69115 Heidelberg, Germany. If you have any concerns regarding our products, please contact ProductSafety@springernature.com

Printed and bound by CPI Group (UK) Ltd, Croydon, CR0 4YY
01/10/2026
02225910-0012